WILDER

Some other titles in the Bloomsbury Sigma series:

Sex on Earth by Jules Howard
Spirals in Time by Helen Scales
A is for Arsenic by Kathryn Harkup
Suspicious Minds by Rob Brotherton
Herding Hemingway's Cats by Kat Arney
The Tyrannosaur Chronicles by David Hone
Soccermatics by David Sumpter
Wonders Beyond Numbers by Johnny Ball
The Planet Factory by Elizabeth Tasker
Seeds of Science by Mark Lynas
The Science of Sin by Jack Lewis
Turned On by Kate Devlin
We Need to Talk About Love by Laura Mucha
Borrowed Time by Sue Armstrong
The Vinyl Frontier by Jonathan Scott
Clearing the Air by Tim Smedley
The Contact Paradox by Keith Cooper
Life Changing by Helen Pilcher
Sway by Pragya Agarwal
Kindred by Rebecca Wragg Sykes
Our Only Home by His Holiness The Dalai Lama
First Light by Emma Chapman
Models of the Mind by Grace Lindsay
The Brilliant Abyss by Helen Scales
Overloaded by Ginny Smith
Handmade by Anna Ploszajski
Beasts Before Us by Elsa Panciroli
Our Biggest Experiment by Alice Bell
Worlds in Shadow by Patrick Nunn
Aesop's Animals by Jo Wimpenny
Fire and Ice by Natalie Starkey
Sticky by Laurie Winkless
Racing Green by Kit Chapman

WILDER

How Rewilding is Transforming
Conservation and Changing
the World

Millie Kerr

BLOOMSBURY SIGMA
LONDON · OXFORD · NEW YORK · NEW DELHI · SYDNEY

BLOOMSBURY SIGMA
Bloomsbury Publishing Plc
50 Bedford Square, London, WC1B 3DP, UK
29 Earlsfort Terrace, Dublin 2, Ireland

BLOOMSBURY, BLOOMSBURY SIGMA and the Bloomsbury Sigma logo
are trademarks of Bloomsbury Publishing Plc

First published in the United Kingdom in 2022

A catalogue record for this book is available from the British Library

Library of Congress Cataloguing-in-Publication data has been applied for

ISBN: HB: 978-1-4729-9038-9 eBook: 978-1-4729-9040-2

2 4 6 8 10 9 7 5 3 1

Illustrations by Tiffany Francis-Baker

Typeset by Deanta Global Publishing Services, Chennai, India
Printed and bound in Great Britain by CPI Group (UK) Ltd, Croydon CR0 4YY

Bloomsbury Sigma, Book Seventy-Four

To find out more about our authors and books visit www.bloomsbury.com
and sign up for our newsletters

For Richard Henry and the countless unsung
heroes of conservation

Contents

Author's Note

When I developed the idea for *Wilder*, I intended to visit every project appearing in its pages. Doing so would have allowed me to describe the innovative work of global rewilders using first-person anecdotes and rich sensory details, and on-the-ground reporting would have brought me into contact with the people responsible for making rewilding a reality. Moreover, I'm accustomed to reporting on initiatives that I have witnessed with my own eyes. It's how I work, how I write.

Covid-19 had other plans.

The pandemic began disrupting life in the UK a week after I submitted my book proposal, and Bloomsbury commissioned *Wilder* several months later. Although lockdowns and travel restrictions were in place by the time I put pen to paper, I continued to believe, optimistically, that restrictions would lift before my deadline arrived: I would be able to travel, not to every destination, perhaps, but to a number of project sites.

Instead, the pandemic lingered, so I changed gears. I homed in on places I had visited in the past and spent countless hours twisting my brain in knots as I considered how to write compelling narratives about places I haven't seen and people I haven't met. Writers of fiction are used to going on journeys of the mind, but writing fiction isn't my strong suit. Besides, journalists don't invent stories – they build them. Like physical structures, stories require robust construction materials, from identifiable characters to narrative arcs.

For me, sketching chapters about unfamiliar places proved a colossal challenge, but something shifted when I started work on Chapter Three, which examines invasive species eradication in New Zealand. I read about a forgotten conservation pioneer named Richard Henry who died nearly 100 years ago (in 1929). As far as I was concerned, Henry was the story's clear protagonist, but had I been able to visit New Zealand while writing *Wilder* I would have paid greater attention to modern-day rewilders. Henry would have become a bit character, not a star, and I am certain that would have been a mistake.

Writing about Henry's quest to save New Zealand's flightless birds gave me the chance to experience a new kind of travel, one that doesn't require aeroplanes or interviews. I realised that even topical stories like those shared in the coming pages can – and perhaps should – incorporate historic characters and events. While I refined the art of reporting from afar, we all learned to live in a smaller, less mobile world, if only for a time.

Perhaps you are asking yourself why I charged ahead. Why not postpone *Wilder* until I had the chance to travel as intended? There are several answers to that question, but the simplest goes something like this: as a concept and practice, rewilding is experiencing growth so exponential, journalists like me can barely keep up. I knew from the outset that I wanted to focus my attention on current rewilding initiatives, so postponing this book would have meant setting certain projects aside, instead identifying new ones if and when life returned to 'normal' (will it ever?). Besides, international travel comes with a hefty environmental price tag. If I could reduce my carbon footprint *while* honouring my vision for this book, I was in.

The rewilding efforts detailed in the coming chapters are all current, but that doesn't mean they are complete. On the contrary, they are ongoing and, in some cases, embryonic. As a result, some projects will have changed or expanded between the time *Wilder* was completed and published – a period of approximately 10 months. Of course, all authors writing about contemporary events face a similar dilemma: how does one finish telling a story when its actual ending has not been reached?

Although I relish neatly ending stories, wrapping them up in silky red bows, I am forced to acknowledge that narratives about rewilding never come to a hard stop. Like nature itself, they – and the projects they describe – are constantly evolving.

Introduction

'We must not go quietly into this impoverished future.'

Ripple, W. *et al.* (2016)

I slipped into Sudan's enclosure at Kenya's Ol Pejeta Conservancy and cautiously approached the gentle giant. Then aged 43, the planet's last male northern white rhino was sleeping on his side. I hesitated, frozen in the background, until one of his round-the-clock keepers motioned for me to approach. Standing beside Sudan, I placed my hand on his ribcage and watched as it floated up and down in seeming slow motion.

Feeling self-conscious but compelled, I quietly apologised to Sudan for what our species did to his. My words, of course, weren't for him: they were for me. Being in his presence felt like visiting a loved one's deathbed – something I've only done once and not very well – but with Sudan, guilt overlaid sadness. I knew that his passing (which occurred two years later) could spell the end for northern white rhinos – and people were to blame.

Meeting Sudan was a privilege and so much more. I am rarely at a loss for words, but there is no way to adequately describe how I felt that afternoon. Words in this case, are not enough. One sentiment, however, is easily conveyed: seeing Sudan was like staring extinction in the face. I doubt I will have this privilege – or punishment, depending on your perspective – again, so I resisted the urge to push discomfort away.

It never left me, but discomfort was soon joined by a wisp of hope. Beside Sudan's enclosure was a larger one containing a number of endangered animals including the last-surviving female northern white rhinos, who were contentedly lounging in the sun like a pair of house cats. Like Sudan, they seemed completely unfazed by their celebrity status and the immense pressure that rests on their humped shoulders.

In the coming days and weeks, I interviewed scientists working to prevent the northern white rhino's extinction. They detailed a bold proposal involving invitro fertilisation, surrogate southern white mothers and years of captive breeding. If their radical plan works, northern whites (or northern/southern hybrids) may return to Central Africa one day. I wrote a *Popular Science* magazine article about the initiative and considered exploring it further in this book, but my publisher wasn't convinced that devoting a chapter to the planet's most imperilled rhino made sense.

In his opinion, the project, at least as it stands, does not qualify as rewilding. I wasn't so sure: rewilding takes time – decades if not centuries – and should, therefore, be viewed on a spectrum. Just because the northern white rhino captive breeding programme is in its infancy does not exclude it from rewilding's purview. The projects featured in the coming chapters all started somewhere, which begs the question: at what point in time does a rewilding plan become a project? As a former lawyer and someone who prizes clarity, I would love to answer this question for you, but when I transitioned from law to journalism 12 years ago, I shifted roles. I would no longer provide answers; instead, I would raise questions. After all, I am not a biologist nor a conservation manager. My sphere is storytelling, and what are stories if not invitations to learn, reflect and develop opinions of one's own?

There was another reason for steering clear of northern white rhino rewilding, one on which my publisher and I ultimately agreed. Although we were determined to select programmes that highlight the boldness of rewilding (and the hope it inspires), we knew we had to draw a line between projects that are in motion and those that may never lift off. With numerous rewilding initiatives gaining ground every day, it was important that I focus my attention on efforts that have a high probability of success. Besides, the most radical rewilding projects – like proposals to return cheetahs to North America or resurrect Siberia's woolly mammoths – will always get airtime. Less likely to make headlines are freshwater mussel reintroductions in San Antonio, Texas, or tales of long-forgotten conservation pioneers (Chapters Five and Three, respectively). Which is not to say that *Wilder* avoids high-profile rewilding initiatives. Indeed, several appear in this book. My goal is to present an eclectic mix of projects that demonstrate the many forms rewilding can take. Before we learn about those forms, however, we need to talk about rewilding itself.

What is it? How does it differ from conservation and restoration? And can it really change the world, as the title of this book suggests?

A group of American conservationists coined the term in the 1990s while calling for a paradigm shift in conservation. In their opinion, large wilderness areas had traditionally been protected because of aesthetic value and moral obligation. 'The scientific foundation for wilderness protection,' they argued, 'was yet to be established.'[1] Rewilding was the solution. Neatly summarised, rewilding came with a nifty acronym: Three Cs that stand for 'carnivores', 'cores' and 'corridors'. In a nutshell, pioneers of the practice advocated for protecting large swathes of wilderness cores while maintaining (or creating) connections between them – whether through wildlife corridors or what are called 'stepping stones'. The third tenet involved the reintroduction of apex predators to areas where they had vanished. Reintroducing carnivores – like grey wolves to Yellowstone National Park in the US Midwest in 1995 – is critical, because apex predators regulate ecosystems. They create what are called trophic cascades, which, in layman's terms, represent the impact predators have on prey as well as the resulting effects that ripple through the entire food web. As these effects permeate, ecosystems begin to bounce back to a healthy, natural state. Rewilding that prioritises the re-establishment of these cascades is known as trophic rewilding.

Rewilding gained traction as scientists advocated for 'novel, process-oriented approaches to restoring ecosystems'.[2] Given the state of our environment, bold approaches and targets are desperately needed more than ever before. We are witnessing the sixth great extinction, the only extinction event caused by human impacts, and we're experiencing unprecedented climate change (while writing this book, the International Panel on Climate Change (IPCC) released a damning report on the state of our planet – the UN secretary-general called it a 'code red for humanity'). Within conservation, rewilding came to be seen as a radical practice with incredible potential. It didn't take long for the theory to expand in scope as academics and practitioners considered its application.

Before long, the 'C' representing 'carnivores' became a 'K' for 'keystone species'. Such species are critical: they are the glue holding ecosystems together and, because they influence environments more than other organisms, their absence can cause devastating domino effects. Keystone species are often, but not always, predators. In some places, large herbivores exert as much influence on ecosystems as their carnivorous counterparts, and let's not forget that in some parts of the world – the Galápagos Islands, for instance – large carnivores are not native. Besides, ecosystem engineers, like beavers and sea otters, are critically important to healthy ecosystems, and animals lower down on the food chain can play important roles, too, by offering bottom-up (versus top-down) services, as we see in Chapter Five. As it turns out, the pyramid-shaped food chain we learned about in school is somewhat misleading.

Another, more extreme shift occurred on the other side of the Atlantic. In the UK and Europe, rewilding experts predominantly advocate for a more passive form of rewilding. Human-led species reintroductions have occurred – most recently of beavers to the UK – but 'passive rewilding' has become the dominant approach in this part of the world. It typically takes place on former agricultural land and is driven by changes in industry and lifestyle. As people abandon agriculture in droves, landscapes have the chance to become wild again. If Yellowstone's wolves are the prime example of active (or trophic) rewilding, the passive approach has Chernobyl as its poster child (it's no surprise that David Attenborough's 2020 documentary, *A Life on Our Planet*, opened near the former nuclear power plant). Passive rewilding promotes a hands-off approach that allows nature to find its own way. As a result, historic baselines – such as returning an environment to its pre-Industrial Revolution state – are typically irrelevant. Passive rewilding doesn't pursue wildness of the past; it embraces wildness of the future.

In recent years, rewilding has further expanded in form and scope while gaining popularity. Like a firecracker sending sparks every which way, rewilding has splintered in numerous directions. You can now rewild yourself, your garden and, if PR professionals adopt the term, your closet.

After seeing the term proliferate, several conservation organisations strategically incorporated rewilding into their branding. One went so far as changing its name – from Global Wildlife Conservation to Re:wild – while countless others brought their conservation work under rewilding's umbrella. For some practitioners, rewilding's evolution has caused discord. Debates over the principles underpinning the philosophy have exploded, with academics bemoaning the term's loose application. As the authors of a 2019 journal article pointed out, 'There are currently a dozen definitions of rewilding that include Pleistocene rewilding, island rewilding, trophic rewilding, functional rewilding and passive rewilding, and these remain fuzzy, lack clarity and, hence, hinder scientific discourse.'[3] Another paper, also published in 2019, argued that the 'diversity of perspectives raises the question of whether there is any common thread within rewilding.'[4] To resolve these ambiguities, several scientists have asked the International Union for Conservation of Nature (IUCN) to resolve the debate once and for all by publishing a clear definition and guiding principles. The IUCN accepted the challenge in September 2021 but is unlikely to deliver anything concrete before its next global congress – in 2024.

As a lawyer and wordsmith, I appreciate concerns surrounding the term's definition: if rewilding means everything, does it mean anything? And how can rewilders synchronise their efforts to ensure the greatest possible impact if they work from disparate assumptions and guidelines? On the other hand, I can't help but roll my eyes at the antics of academics in ivory towers. They remind me of two bickering siblings, each begging their parents to discipline the other. While they bicker, conservation ticks along.

Linguistic sticklers seem to have forgotten that rewilding is happening all over the world with increasing frequency – sometimes with the term attached, sometimes without it[*]. On the other hand, people managing rewilding projects are far less concerned with the term's meaning than they are with the work itself. I interviewed dozens of them while

---

[*] In some cases, rewilders fail to use the term when describing their work because they are unfamiliar with it. Other times, conservationists intentionally avoid mentioning rewilding because people in the region where they operate possess negative perceptions of rewilding. British farmers, for instance, sometimes see rewilding as agriculture's enemy.

writing this book. One – Mark Stanley Price of Chapter Two – addressed the debate in a particularly astute way, remarking that, 'Rewilding is hugely appealing as a concept, but we must acknowledge that it means different things to different people.' Mark spoke about the critics who want to abandon the term on the basis that it's convoluted by too many definitions. 'I disagree with that fundamentally,' he said. He embraces a more-the-merrier approach.

Rewilding's malleability could be a powerful thing, asserted scientist Dolly Jørgensen in 2014. The term's plasticity, she wrote, allows it to 'capture the public imagination' and cross between political and scientific discourses. The risk, of course, is that 'rewilding becomes a word "full of sound and fury, signifying nothing" – or perhaps, signifying everything'.[5]

When I began developing the idea for Wilder, I didn't have a working definition of rewilding. I had never heard of the Three Cs and was naive to debates raging in the halls of academia. And yet, I knew what the concept meant: I understood it implicitly. Rewilding involves restoring some measure of 'wildness' to places humans have altered and destroyed. Never mind that 'wild' (and 'wilderness') mean different things to different people. I'd know rewilding when I saw it. Although I understood that rewilding was a form of conservation, I observed a clear distinction between the two practices: if conservation seeks to maintain what is left and stave off further declines, rewilding goes a step further by attempting to revive entire ecosystems and the species they lost. Conservation and rewilding work in concert, of course, and are often conflated, but the bottom line is that rewilding aims higher. For some, resurrecting extinct species is the ultimate form of rewilding. If 'Pleistocene rewilders' have their way, woolly mammoths will roam the plains again one day.

Once the book was commissioned, I had no choice but to tumble down the rewilding rabbit hole. I read the academic literature on the subject, interviewed leading experts – including several of the scientists who coined the term – and engaged my conservation peers in debate about its meaning. The deeper I went into the metaphorical rabbit hole, the harder it was to claw my way out. For months, I was paralysed by a self-imposed compulsion to unlock the riddle of rewilding. I was thinking like a lawyer instead of a journalist, intent on answering (instead of raising) questions.

Once I realised my mistake, I corrected course. I temporarily put linguistic and philosophical questions aside as I began the far more interesting task of considering which rewilding projects to highlight. There are hundreds – perhaps thousands – in existence, some large, some modest. I already had several in mind, but numerous chapters were up for grabs. Where to begin?

As a supporter of the emerging 'conservation optimism' movement, I knew I wanted to showcase novel, highly ambitious rewilding initiatives. Doing so meant focusing on active trophic rewilding projects built around the Three Cs. A line had to be drawn, however, between ambitious-yet-feasible projects and efforts that are unlikely to succeed (or are years away from launching). I was intent on taking a global perspective, and wanted to pay particular attention to areas where rewilding has yet to become mainstream, which explains why nearly all of the initiatives I spotlight occur outside of Europe and North America. Likewise, conversations about rewilding have focused on the Western world despite the fact that some of the most critical and innovative initiatives are happening elsewhere. In the interest of inclusivity, I initially hoped to write about rewilding on every inhabited continent, but pandemic-induced lockdowns hindered my ability to travel, which in turn reduced my ability to gather information about places I have never visited. With that in mind, I decided to prioritise places I have seen with my own eyes.

By incorporating familiar places, I had the opportunity to simultaneously share part of my personal story with you. Snippets scattered throughout the coming pages paint a picture of a person who began life with an innate zest for nature. Unfortunately, my enthusiasm for nature and wildlife dimmed as I grew up. Human dewilding happens, I suspect, to most of us as we cast off our childlike wonder in pursuit of that amorphous thing called adulthood. If only we had listened to Peter Pan. As you will see in the chapters that follow, I eventually reclaimed my wonder for the natural world while shedding the perceived expectations family and society had placed on me. I came to conservation (somewhat belatedly) full of curiosity and passion. And, let's be honest, fear – for the future generally as well as the vulnerable species (like northern white rhinos) that may not be around for much longer.

Facing this uncertain future, it's only natural that we ask ourselves how *we* can help save this beautiful, complex, damaged planet of ours.

That is another question I can raise but not answer, but I can tell you this: rewilding ourselves and our surroundings is a great place to start. David Attenborough goes further by calling rewilding our *only* solution. Midway through *A Life on Our Planet*, Attenborough says: 'To restore stability to our planet, we must restore its biodiversity – the very thing that we've removed. It's the only way out of this crisis we have created. We must rewild the world.'

CHAPTER ONE

A Park for the People

On a grainy 1960s travel video, a lioness and her cub rest in the decrepit entryway of Gorongosa National Park's famed Lion House. The structure became a favourite hangout among the Mozambican park's prides after severe flooding forced staff to abandon it years earlier. At the edge of the small concrete house, another female navigates a spiral staircase leading to the roof – a useful, if unnatural, lookout point.

In those days, decades before the Mozambican Civil War ravaged the country and Gorongosa, travellers flocked to the million-acre (3,674km²) park located in East Africa's Great Rift Valley. Marketed as the 'the place where Noah left his Ark', Gorongosa was known for its resident lions. With an estimated lion population of 200, the national park attracted safari-goers from all over the world. American astronaut Charles Duke called his Gorongosa visit 'as thrilling as landing on the moon'[1]; and a Zimbabwean writer said she could have leant out of her car and scooped up a lion cub in her hands. Lions weren't the only animals roaming the park in large numbers: in 1972, ecologist Ken Tinley led Gorongosa's first ever aerial wildlife survey and counted 2,200 elephants, 14,000 buffalo and thousands of ungulates, from zebra to waterbuck and eland. The overall biomass was on a par with famous African wildlife areas like the Serengeti and Ngorongoro Crater. Some believe the park boasted greater biodiversity than any place in Africa.

South African conservationist and pilot Paul Dutton helped conduct the 1972 count. Years later, he published a memoir, *Spirit of the Wilderness*, in which he recalled flying over large swathes of pristine wilderness. Long, exhilarating days were spent in the skies, Paul's beloved aircraft solely guided by a compass and the moving sun. At the end of each day, Paul felt 'thoroughly exhausted, part deaf, but ecstatic with the numbers and variety of animals' he and his colleagues spotted in a range of habitats, from wetlands to savanna. The ecosystem was, he wrote, beyond compare. By capturing rain and feeding the park via several rivers, nearby Mount Gorongosa was the park's 'vital heart', its lifeblood. Little did Paul know that decades of conflict would soon descend upon Gorongosa, stripping away its magnificent biodiversity and leaving in its wake more landmines than lions.

For Paul, fellow white Africans and international travellers, the final decades of the colonial era, which began in 1505, felt like halcyon days. Tourists looking for a European twist on the continent came to Mozambique, often arriving by ship into the port town of Beira before travelling by light aircraft to Gorongosa. Day-trippers toured the park before enjoying lazy, wine-fuelled lunches; others checked into Chitengo Camp, which by 1965 accommodated 100 overnight guests and boasted two swimming pools.

Romantic for some, the colonial period was tainted by racial inequity. While Portuguese Mozambicans and white foreigners soaked up Gorongosa's splendour, black Mozambicans were prevented from entering the park – unless they were there to serve chilled beer to guests while sporting crisp khaki uniforms. But although locals like Roberto Zolho – who studied in the park in 1981 and later served as park warden – weren't allowed inside, many knew about the ecological hotspot that has been protected since the 1920s when part of it was set aside as a hunting concession (a common start for many African protected areas). Conservation as we know it began when big game hunters realised the sport would dry up if they didn't preserve their living quarry for future hunts*. Indeed, some of the most famous conservationists hunted for sport. US President Theodore Roosevelt, who is often credited with creating the US National Park system, a model that in turn inspired protected areas around the world, killed thousands of African animals during a 1909–10 expedition. In his defence, the spree wasn't *just* about collecting trophies: working with the Smithsonian Institution, Roosevelt's team helped build the institution's vast collection of animal specimens.

The scales of racial injustice began to shift when, after nearly 500 years of colonial rule, Mozambique gained independence from Portugal in 1975. For the first time in modern history, Mozambicans began to understand that Gorongosa and their country's other natural treasures belonged to them, the people. However, independence was costly. Tens of thousands of Mozambicans were killed or tortured during the decade-long War of Independence (1964-74), and up to a million were forced into relocation camps.

Gorongosa, on the other hand, emerged virtually unscathed. A 1976 survey counted thousands of elephants and the park's largest-recorded population of lions.

---

* The African bluebuck, which lived in South Africa and vanished circa 1800, was the first large African mammal to go extinct during modern times. Historian Nigel Rothfels, whom we meet in the next chapter, likens the disappearance of African species like the bluebuck and quagga to the decimation of bison across the United States. Losses like these prompted concern 'among both big-game hunters and zoo proponents'. 'Importantly,' writes Rothfels, 'these hunters and zoo advocates were often one and the same people.'

As Portuguese families fled the country, a new communist government led by Frelimo (*Frente de Libertação de Moçambique*) set about creating opportunities for black Mozambicans like Roberto Zolho. Donor funding financed training courses at Gorongosa – programmes Dutton helped establish and run – and the South African soon found himself basking in what he calls 'euphoric days of post-independence'.

Initial programmes trained rangers. Then, in 1981, Gorongosa hosted its first ever protected-area management course. A 21-year-old forestry student at the time, Zolho was one of nine Mozambicans recruited to join the 10-month course based inside the park. Zolho and his classmates arrived in May 1981, an experience he remembers fondly. Wildlife was thriving, both in terms of species diversity and sheer numbers, and the cohort was eager to explore the lush ecosystem that had, until then, been inaccessible to them. However, their enthusiasm was hampered by concerns about the spread of another war, this one civil. Frelimo government forces were battling Renamo (*Resistência Nacional Moçambicana*) – a rebel group that was backed, and arguably created, by South Africa and Zimbabwe (both countries' governments abhorred the idea of a communist neighbour). The 1977–92 Mozambican Civil War was concentrated in the countryside, with much of the fighting in the centre of the country, dangerously close to Gorongosa. 'We knew the war was all around us,' says Roberto. 'The rebels, the Mozambican National Resistance, or Renamo, had their base in the nearby mountains'.

Park tourism slowed down as conflict escalated (an attack in 1973 during the struggle for independence had already put a dent in it), but Gorongosa otherwise carried on operating normally. During the day, Roberto and his classmates sank their teeth into practical and theoretical lessons that touched on everything from anti-poaching to ecology. Sessions were led by visiting lecturers and three primary teachers from Chile, Britain and Tanzania.

Then, on the morning of 17 December 1981, the course came to an abrupt and perilous end. The balmy Thursday began uneventfully. Paul, his fellow teachers and their students sat in a makeshift classroom on the veranda of Chitengo Camp where 15 years earlier safari-goers sipped Portuguese wine while reflecting on magnificent game drives. 'We were in the middle of class,' remarked Roberto, 'when something alarming happened.' From their perch on the veranda of Chitengo Camp, Roberto and his classmates watched in horror as a truck that had recently left camp to fetch firewood from outside the park returned much too soon. In it were people wearing government uniforms.

Within minutes, Roberto's concerns morphed into panic as his Tanzanian teacher bolted across the veranda shouting that rebels had arrived. The teacher jumped a nearby fence; Roberto and his classmates quickly followed. Powerful instincts, the very ones that keep impala alert to big cats on the hunt, led them to safety, but the group had also mentally prepared for the possibility of a rebel-led siege. Feet thumping through uneven bush, the group ventured east, running for six hours until reaching a town called Muanza on Gorongosa's eastern edge. They could have exited the park faster by going in the other direction, but with rebels entering from the western gate doing so would have been a catastrophic, even deadly, mistake. From Muanza, the government organised safe transport to Beira, but it soon became clear that the programme couldn't safely continue in Mozambique, so the government arranged for all nine students to attend a similar course in Tanzania.

Paul also escaped the skirmish – in his trusty plane – but Renamo forces murdered several members of staff before taking the English and Chilean teachers, John Burlison and Moisés Carril, hostage. The two men were accused of working for the government's secret police – an absurd, unfounded claim – but were released in 1982.

In the months that followed, violence and poaching spread like wildfire through Gorongosa's savannas and swamps as Renamo established headquarters near the park. Like wilderness areas in other war-torn regions, the national park and nearby Mount Gorongosa offered terrain and seemingly infinite natural resources that together lent themselves to guerrilla warfare. Gorongosa's central location within the country made it that much more desirable. During the 16-year-long civil war, bloody battles were fought inside the park, and at various points during the conflict both sides claimed control of Gorongosa. Within a year or two of the December 1981 attack, the park officially closed, but several rangers reportedly remained behind to carry out their duties. Details are murky, but three are said to have died.

Roads and structures, including Chitengo Camp, were destroyed by aerial bombings. Animals were killed for sustenance and, depending on their commercial value, to fund the war. Gorongosa's elephant population plummeted from 2,200 to 100, with many killed for their tusks, the sales of which financed the purchase of supplies and weapons. Hooved animals like zebras, wildebeest and buffalo, all of which were seen as easy protein, fared even worse: fighters killed them in the thousands. As their numbers dwindled, lion populations did too. Some lions were also gunned down for sport. It was a poaching free-for-all. Meanwhile, members of communities located near the park's boundaries were killed or forced by military forces to abandon their homes and move to makeshift villages miles away.

Over the course of 16 years, the war took a devastating toll on the East African nation: up to 1 million people died and another 5 million were displaced. Although the worst violence occurred in the south of the country, communities living near Gorongosa suffered immensely. As time wore on, access to food supplies dried up, so locals turned to the park's wild bounty to survive. By the time peace arrived in 1992, most of Gorongosa's large mammals had vanished, their wild territories overtaken by thousands of people now living within the park's 400,000 hectares (988,422 acres). A 2000 survey counted around 20,000 park residents. Without proper park management in place, unsustainable hunting continued for several years, the impacts of which were only measured in 1994 when Paul flew to Gorongosa to conduct the first post-war wildlife survey.

His homecoming was joyless. 'I returned … to a park that showed only the scattered, bleached bones of the once abundant wildlife, sacrificed to feed a war without victors,' wrote Paul in his memoir. 'The camp had been reduced to rubble, leaving a habitat suitable only for black mambas and Mozambique spitting cobras.' Unlike blissful surveys of the past, this one demonstrated that most large mammal populations, including all of the so-called 'Big Five' (elephant, buffalo, rhino, leopard and lion), had shrunk in number by 90 per cent or more. Paul didn't observe any hippos, wildebeest or zebras, nor did he see a single lion. He counted a meagre 100 elephants; and Gorongosa's buffalo herd, which once numbered 14,000, appeared to have vanished into thin air. Years later, it became clear that buffalo, hippos, lions, wildebeest and zebras hung on by a thread, but African wild dogs, black and white rhinos, spotted hyenas and leopards were wiped out. All told, the Mozambican Civil War reduced the park's total biomass by at least 93 per cent.

Birds, small critters and flora fared better than large mammals, but declines in plant-eating animals caused dramatic changes to Gorongosa's vegetation. Landmines littered the park's vast acreage and many more lurked on nearby Mount Gorongosa. They were imperceptible, a silent, unseen threat not unlike a tiger awaiting its prey. Assaults were evident among a scarred populace, and years later Gorongosa's current park warden, Pedro Muagura, witnessed the death of a young boy who stepped on a mine. Long before he accepted this (or any full-time) position at Gorongosa, Pedro began spending time in the war-torn park.

Like Roberto, he learned about the park as a boy but couldn't pass through its gates because of the colour of his skin. When he finally arrived in 1992 – after the civil war, to teach forestry students – he witnessed a landscape devoid of the biodiversity that piqued his young mind's interest. 'It took days to see a single baboon,' Pedro tells me, 'and up to five to encounter a warthog. My students and I never saw elephants or lions even though we spent several weeks camping in the park.' His students were so disappointed by the lack of animals and what it meant for a career in conservation that they tried to drop his class, but the government recognised Mozambique's potential as a wildlife destination, and required all secondary school students to take forestry and wildlife management courses. Pedro tried inspiring his students by showing them field guides and explaining that, although annihilated, Gorongosa would thrive again – in due course. But they, like many others, were sceptical. Who could blame them? War was all they they'd ever known.

Thirteen years after running for his life through the wilds of Gorongosa, Roberto was hired as the park's first post-war warden. Like Paul, he despaired at the state of the land once called the 'Eden of Africa'. Although no one dared to imagine a fully fledged comeback, small steps towards rewilding commenced in 1994 when the African Development Bank (ADB), European Union and International Union for the Conservation of Nature (IUCN) provided funds for a five-year rehabilitation project.

Urgent priorities included removing landmines, restoring roads and other park infrastructure, putting a halt to illegal hunting and beefing up park staff. On the government's recommendation, Roberto hired ex-soldiers from both sides of the war to work as park scouts. The militiamen knew the park inside and out, making them invaluable to recovery efforts, especially those involving demining and anti-poaching. Although tension between wartime enemies never fully dissipated – conflict between Renamo and Frelimo still occasionally erupts and is one reason Gorongosa has struggled to attract tourists – Roberto says his cadre was willing to move forward peaceably.

Newly minted scouts spent hours driving through the park, often without seeing a single animal. Paul went on long drives as well, confiscating weapons from former soldiers. Hunting and poaching continued during this post-war period; by some accounts, it escalated as locals sought to feed and support their families. In 1994, the IUCN estimated that upwards of 60 tonnes of animal carcasses were removed from Gorongosa every month, with much of the meat sold in nearby towns (experts like Pedro question this estimate, as do I given the aforementioned carnage).[2] The park was, as a 2016 magazine article put it, 'broken ... entire chains in the food web had been and remained decimated'.[3]

Years of war and unsustainable resource extraction also altered Gorongosa's vegetation. Grasses grew tall, tree cover expanded and an invasive shrub, no longer kept in check by elephants and other browsers, took root. In the late 1990s, scientists were particularly worried about the park's two most important habitats: savanna and the montane rainforest of Mount Gorongosa – which became part of the national park in 2010. The 1,800m (5,905ft) peak, which sits at the southern edge of the Great Rift Valley, contains exceptional biodiversity. It is home to numerous endemic species – those not found anywhere else in the world – including a newly discovered bat. Even though some of the best minds in science – among them famed biologist E. O. Wilson – have surveyed its terrain, much of the mountain's biodiversity remains undiscovered.

Mount Gorongosa's importance, however, was never in dispute. In addition to hydrating the park via numerous rivers and streams, the mountain is home to roughly 500 tree species that protect people and the environment by producing oxygen, storing carbon and absorbing rain, thus preventing soil erosion. Important as it was, the mountain suffered severe losses as locals felled trees and converted forested slopes into farmland; thousands lived on the mountain, too, where they planted crops like beans, maize and potatoes. In 2007, scientists estimated that the mountain's ecosystem needed urgent protection and reforestation; otherwise, within half a decade it would degrade beyond repair.[4]

Day by day, Roberto and his colleagues made headway, and soon animals started showing their faces and rumps again. Some species experienced natural rebounds – in the case of certain antelope, like waterbuck, the predator void meant numbers had nowhere to go but up. In fact, Gorongosa's waterbuck population has arguably grown to excess with numbers now beyond historic levels. In other cases, highly intelligent animals like elephants only reappeared when conditions seemed sufficiently safe.

In 1997, a foreign journalist remarked in the *Christian Science Monitor*, 'While Gorongosa may be devastated, experts say it will regenerate because it's being rescued in time.' But the same article spoke of poachers armed with AK-47s, crumbling walls and a countryside containing up to a million landmines that maimed or killed 40–50 people each month. The article also stated, somewhat presciently, that 20 years would pass before wild animals returned to pre-war levels.

Glimmers of hope helped weary conservationists return to work each day, and signs of wildlife meant entities like the African Development Bank had reason to invest in Gorongosa's future. However, the initial restoration project launched in 1994 ended in 1999, leaving park staff worried that Gorongosa's restoration would stall – perhaps indefinitely. 'We'd established enough infrastructure to keep the park going, but needed more resources to scale things up,' explains Roberto. For example, by 2000, the park's boundary had been secured and roughly 20 ranger posts were in place, but communities continued living in the park and turning to its natural resources for sustenance. As evidenced by Paul's aerial survey and ongoing counts, large mammals had experienced catastrophic declines. Some, like buffalo – once the most populous large mammal in the park – would never bounce back without targeted, human-led interventions. Put another way, Gorongosa desperately needed active trophic rewilding. But who would pay for it?

September 2004

On a cool autumn day in Boston, Massachusetts, 12,553km (7,800 miles) from Gorongosa, entrepreneur and philanthropist Greg Carr greeted Mozambique's then-president, Joaquim Chissano, inside Harvard's John F. Kennedy School of Government. Carr, a graduate of the institution, invited Chissano to give a 2004 lecture after developing an interest in Mozambique several years earlier.

His interest wasn't merely intellectual: flush with cash following several successful tech ventures, the multimillionaire had recently decided to walk away from the tech industry (and all for-profit activities) to dedicate himself to humanitarian activities. In 1999, he endowed the Carr Center for Human Rights Policy at Harvard while creating his own foundation, whose focuses include the arts, the environment and human rights. Having established the Carr Foundation, he began the momentous task of considering where to allocate his funds and energy.

He wanted a single, long-term project that he could engage with for the rest of his life – and his sights were set on Africa. After a neighbour introduced him to the Mozambican ambassador to the United States, Carr spent two years studying the country while building relationships. Carr is not a parachute-in kind of philanthropist: When he commits to something, it's for the long term. Having carefully studied Mozambique, he arranged to visit six of the country's wilderness areas, including Gorongosa, during the spring of 2004.

Roberto Zolho, now in his eighth year as park warden, spent the day with Greg. Like birds of prey surveying the world from the sky, Greg's first views of Gorongosa were from a helicopter. He knew it was 'the one' after his day-long visit but emphasises that years of research allowed him to make a quick decision on the ground. 'I knew from voracious reading that this was one of the most important natural areas in Africa,' he says. 'Notably, although the park didn't contain much wildlife, the ecosystem otherwise seemed intact. Rivers were flowing, the lake was full and forests remained, so the ecosystem's building blocks were all there.' Greg was also keenly aware of the suffering locals had experienced during decades of war, and figured he could help by creating an integrated conservation and development project. He envisioned a healthy park that would lift up communities by creating a green economy.

Seven months later, in October 2004, the Carr Foundation donated $500,000 to the restoration project. A year later, Greg upped the ante by pledging $40 million over the course of 30 years as he and the Mozambican government entered into a memorandum of understanding (MOU) to restore the park (their joint management agreement took more time to develop and wasn't inked until January 2008). After signing the MOU, they commenced the process of bringing Gorongosa back to life by rebuilding infrastructure, fostering economic development opportunities and restoring the park's wildlife.

Greg fell in love with Gorongosa, as evidenced by the fact that he spends half of his time there, but when it comes to management, he has what he calls 'a light touch'. His only professional title? 'Member of the oversight committee'. Yet his financial commitment continues to grow. Every year, he donates a minimum of $5 million to the project, and he's committed until 2043, making his overall pledge $200 million.

For his part, Greg considers the collaboration as much about human development as conservation. This joint aim makes Gorongosa, in his words, a 'human rights park'. The 200,000 or so people living adjacent to the park (in what are called 'buffer zones') require considerable support given that they're among the poorest communities in a country suffering from severe poverty, low life expectancy and high HIV rates. Although some locals resisted restoration, knowing it would prohibit them from extracting resources from Gorongosa, others were able to see the park's economic and tourism potential. It helped that communities were consulted from the get-go: Greg has spent countless hours with local leaders negotiating the nuanced relationship that binds them. Today, he's beloved, a local celebrity, but there were plenty of stumbling blocks along the way. When Greg visited one of the local communities for the first time, he upset locals by arriving in a red helicopter (in their culture, the colour red signifies angry spirits). Once he began communicating his vision for the region, he discovered that his motives were often in question.

'People see a smart foreign person come in and make a commitment to them, and they can't help but wonder what his angle is,' explains Marc Stalmans, Gorongosa's director of scientific services. 'They were weighing him up because, to them, putting money into this kind of project seemed foolish.' As time passed and advances accrued, local communities began to understand that Carr was in it for the long haul. When it came to human rights and community engagement, he wasn't, as Marc puts it, ticking off boxes (these days, all savvy conservation organisations boast about community development projects in the same way corporations tout sustainability initiatives when, in reality, many are simply greenwashing their activities). In fact, Gorongosa's engagement with nearby communities was lauded by one journal article as 'the first conservation collaboration of its kind in the world'[5]. Greg puts more of his money into initiatives taking place *outside* the park than inside it. Most recently, he promised to fund the building of 60 new primary schools.

In November 2005, a year after the Carr Foundation and Mozambican government signed a MOU to restore Gorongosa, species rewilding commenced in earnest with the construction of a fenced wildlife sanctuary within the park (which isn't itself fenced). The 6,200-hectare (153,201-acre) area would serve as a 'soft release' site – a place where animals translocated from other African regions could safely adjust (and breed) before spreading into the wider ecosystem. Ten months later, Gorongosa welcomed its first shipment of animals: 54 buffalo from South Africa's Kruger National Park. The species was a fitting choice for the park's first phase of animal rewilding given the vast herds that roamed Gorongosa before the civil war.

However, Gorongosa's rewilding efforts didn't begin with buffalo for symbolic reasons. Reintroductions involve translocating animals between regions, which is complex and risky. When they occur (and in what order) depends on ecological health and inter-species dynamics. There's no point in returning large carnivores to an area without sufficient prey; conversely, reintroducing large numbers of prey species without ample predators to control their growth brings a different set of risks – from overgrazing that causes unsustainable vegetation loss to overpopulated herds that starve in the absence of ample food. The latter outcome has cast serious doubt on the success of Europe's most famous (or infamous, depending on whom you ask) rewilding experiment: the Oostvaardersplassen nature reserve in the Netherlands, where large herbivores have been reintroduced without predator counterparts. Unfettered growth has caused thousands of animals to die during harsh European winters; rewilded herbivores either starve to death or are shot by forest rangers before they naturally perish.

The above factors influenced the timing of the buffalo's return – as bulk grazers, buffalo were able to tackle tall, coarse grasses, which in turn opened things up for other grazing species – but Gorongosa's rewilding also hinged on a mundane factor: animal availability. Unlike other species, buffalo were easily sourced while other species posed logistical challenges. Thirty-one additional buffalo relocated from Limpopo National Park joined Gorongosa's herd in 2007; and in the same year 180 wildebeest were brought up from South Africa. A wildlife survey, meanwhile, demonstrated that multiple species were on the rise, and lions, long vacant from Gorongosa's iconic Lion House, were photographed at the site for the first time since the 1960s.

In 2008, the Gorongosa Project officially launched with three core goals: conservation and science; human development; and sustainable tourism. Restoring the park's ecosystem and wildlife to pre-war levels was a significant part of the first aim. Rewilding doesn't always have a clear baseline: some advocates want to return the planet to a past state – say, pre-Industrial Revolution. Passive rewilders, on the other hand, typically ignore baselines. In the absence of robust historic records, scientists working from baselines sometimes have to speculate about the conditions they seek to restore, but the project benefits from the fact that the park has been studied by scientists for decades. A clearly defined end point increases the chance of successful rewilding while ensuring that Gorongosa doesn't become overpopulated – a problem that plagues some wilderness areas.* Species

---

* For instance, elephant numbers swelled in South Africa's Kruger National Park following years of successful conservation, prompting the country's national parks authority to 'manage' the park's population through culls. More than 16,000 elephants were killed between 1966 and 1994 – when the practice ceased. Animal rights activists celebrated the end of the cruel practice, however some conservationists argue for its resumption on the basis that Kruger's elephant population has exceeded natural carrying capacity.

restoration continued with several modest translocations – a tool that involves manually moving animals between regions – that expanded the park's elephant and hippo populations. Concurrently, many of the park's herbivores experienced impressive, if uneven, growth, as evidenced by a 2010 aerial census showing overall wildlife growth of 40 per cent over a three-year span. These results proved that Gorongosa's rewilding was beginning to manifest six years after the Carr Foundation entered the picture.

Marc and his scientific research team were delighted by the census, particularly since much of the growth occurred organically. Elephants and lions rebounded of their own accord, helped by the fact that Gorongosa is what Marc calls 'a really productive environment'. Heavy rains, frequent flooding and fertile soils foster abundant and nutritious plants that sustain the park's herbivores. However, even in a rich environment, certain species bounce back faster than others. Fast-breeding animals, like impala, grow exponentially: in just under three years, numbers will double, says Marc. Conversely, elephants have the longest gestation period of any mammal (up to 680 days), and growth rates are further slowed by the fact that females don't reach sexual maturity until age 10 or older, almost always produce only one calf at a time (twins are exceptionally rare) and don't have another calf for at least four years. All of which explains why the park's herds have only grown from 100 in 1994 to 500 in 2020.

Even after years of restoration, leopards remained absent, with the exception of a lone male spotted in 2018 and 2019 that likely entered the park from a nearby hunting concession. The void was particularly strange given the cat's adaptable nature and large range. While it could be explained by the 200,000-odd people living in Gorongosa's buffer zone, the expert hunter often thrives in human-dominated landscapes. Mumbai, India, is a well-known example. So, why weren't leopards prowling around Gorongosa? Paola Bouley, who leads Gorongosa's carnivore programme, believes the region's leopards succumbed to poaching and snaring. Even after these threats were addressed, leopards did not come back on their own, so Paola and her colleagues vowed to re-establish Gorongosa's leopard population using translocation.

Another mystery involved Gorongosa's lions. Despite a natural rebound, the park's prides were not growing as quickly as anticipated. It wasn't until Gorongosa's carnivore programme began collaring individuals and tracking their movements that the reason for their stalled growth emerged: snares and steel-jaw traps used by locals for bushmeat were inadvertently harming and killing an alarmingly high percentage of the park's biggest cat. Once biologists had this information, they worked closely with wildlife rangers to desnare and better monitor areas where prides spend time – they recovered over 20,000 traps and snares over the course of several years – and within a relatively short period of time pride and coalition size began to increase. As of late 2020, Gorongosa was home to more than 150 lions.

While noteworthy, these triumphs were tempered by environmental degradation on Mount Gorongosa, along whose slopes locals continued to cut down trees and plant crops. When scientists concluded in 2007 that urgent curbing of human activities was required to ensure the mountain's health, local people held a different view. They considered living and farming in the area a moral and legal right – and arguably a matter of necessity. Prohibitions seemed cruel given that communities had already been shut out of the park: first by colonial powers in the 1920s, then by park authorities following the civil war.

Such sentiments are fair and familiar: when it comes to conservation, human rights and environmental protection can conflict. Tension is especially rife when protected areas preclude locals from using the lands of their forefathers, leading some to call the approach 'fortress conservation' and others to denounce environmental regulations as human rights violations. US national parks, for instance, have long served as models for protected areas around the world, but their creation displaced indigenous people. In the words of Native American author David Treuer, '...all of them were founded on land that was once ours, and many were created only after we were removed, forcibly, sometimes by an invading army and other times following a treaty we'd signed under duress'.[6] Treuer now asks that national parklands be returned to Native Americans so the country's indigenous people have the ability to 'tend – and protect and preserve – these favored gardens again'.

For decades, scientists pushed for formal protection of Mount Gorongosa, but Greg had promised a 'human rights park'. Ignoring the interests of locals meant reneging on his pledge, and turning a blind eye to community interests risked losing the support he spent years carefully cultivating. Besides, long-term conservation requires community cooperation and collaboration. Without it, what's the point?

In July 2010, a compromise emerged: the park expanded by 10 per cent (bringing its total size to 406,700 hectares (1,004,978 acres) while the government created a buffer zone of 330,000 hectares (815,448 acres) in which roughly 200,000 local residents are able to live and farm with some government-imposed restrictions. Common in conservation, partially restricted buffer zones create space between protected areas and human communities, and are intended to minimise environmental disturbances and human–wildlife conflict. Most importantly, all portions of Mount Gorongosa above 700m (2,297ft) became part of the national park, ensuring formal protection. With incredible species biodiversity and hydrological value, the highlands needed protection more than low elevation areas, but the latter can't afford neglect.[7] As people who logged or farmed at higher elevations moved their operations closer to home, park staff began ramping up initiatives that would allow locals to continue working at lower elevations without stripping the environment of its natural resources. One solution: shade-grown coffee.

Park warden Pedro Muagura brought the idea to Mozambique after studying agriculture and forestry in Finland and Tanzania, where he obtained two bachelor's degrees. Finland showed him that forests can be protected while providing economic opportunities; and in Tanzania, Pedro learned that coffee plantations can minimise human–wildlife conflict because animals dislike coffee leaves and are therefore unlikely to raid such crops. Coffee farming, he realised, would work nicely on Mount Gorongosa: it offered sustainable employment and encouraged conservation since hardwood trees have to be planted alongside coffee trees to ensure they have adequate shade, making the project as much about reforestation as sustainable farming. It simultaneously creates 'a sense of stewardship', says Gorongosa's director of sustainable development Matt Jordan, because farmers own the land where coffee (and other sustainable products like honey and cashew nuts) are cultivated.[8]

Since launching the programme in 2015, more than 600 community members have planted 200,000 coffee trees (and an additional 100,000 hardwood trees to offer shade) along the mountain's lower slopes – and, as of late 2020, revenues from domestic and international sales of such coffee exceeded US$70,000. By 2025, more than 500,000

coffee trees and 200,000 native hardwood trees will have
been planted across 1,000 hectares (2,471 acres) dedicated to
coffee farming. The initiative further aims to restore 7,082
hectares (17,500 acres) of rainforest by planting 1 million
native hardwood trees by 2035. Wildlife will benefit from
this habitat boost, and reforestation simultaneously mitigates
climate change through carbon sequestration.

While some local community members collaborate with
Gorongosa, others work for the national park. Gorongosa
makes a point of hiring local people whenever possible. As of
April 2021, 95 per cent of park employees were from the
region, with 87 per cent from the nearby buffer zone.
Non-Mozambicans, on the other hand, only represent
0.02 per cent of Gorongosa's staff. Gorongosa also invests in
the country's environmental leadership by hosting a wildlife
conservation master's course inside the park. The first of its
kind, the two-year programme, which launched in 2018,
only admits Mozambicans. It's not as sexy as animal
translocations or new species discoveries, but the master's
course may be the most important piece of the puzzle.

While the Pedro Muaguras and Robert Zolhos of our
story had to leave Mozambique to obtain scientific training,
tomorrow's conservationists are gaining critical skills in
their own backyards, which offers several key advantages.
Because students learn about conservation in the context of
their own country's natural heritage, they understand the
Mozambican conservation landscape before beginning
work. They may also develop attachments to Gorongosa
(and the country's protected area network) that keep them
in-country, thus preventing brain drain.

In April 2018, as students from the inaugural master's
cohort prepared to take their first-year exams, Gorongosa
reintroduced an iconic, endangered species to its vast
terrain. African wild dogs – also known as painted

wolves – had been absent from Gorongosa for 50 years when a pack of 14 reclaimed the landscape. A significant milestone in and of itself, the event marked the first reintroduction of wild dogs in Mozambique. It is part of a wider strategy to stop the decline of a species that has gone extinct in at least 11 African countries (and possibly a further 11[9]) by returning them to areas where they vanished. The pack hunters are wide-ranging and thus less able to cope with habitat fragmentation than species like lions, but, like big cats, they play a crucial ecological role. By keeping prey numbers in check, wild dogs help prevent depletion of native vegetation. As predators, wild dogs and lions can work in concert since the expert hunters tend to target different prey species. Thanks to natural growth and a further introduction in 2019, Gorongosa's painted wolf packs now number 80.

Two and a half years later, in November 2020, Gorongosa brought back another iconic carnivore. Relief washed over Paola and her colleagues when they released a female leopard from South Africa years after initiating the translocation process. In the words of a *New York Times* reporter, the process involved 'three grinding years of begging, bartering, form-filling, false starts and blind alleys'. Overlaying predictable red tape was, of course, the global pandemic. The park's new female, Sena, was joined by three additional leopards in April 2021 and more are on the way. Because Gorongosa's 'founding population' contains males and females, the park won't have to wait long to see wild-born cubs.

In the coming years, spotted hyenas will join them, but several factors – from import permits to the availability of suitable animals – mean that reintroductions don't always happen as quickly as conservationists would like. For instance, Marc and his colleagues have wanted to boost

Gorongosa's zebra herd for some time, but the subspecies native to Gorongosa (Crawshay's zebra) can be hard to source. Rather than import a readily available subspecies from South Africa (and risk diluting the existing population's gene pool), the team opted for a modest translocation of seven Crawshay's zebra from a nearby hunting concession in 2013. Eight more arrived the following year. There is no denying that Gorongosa needed trophic rewilding to restore its ecological balance and improve its prospects as a tourism destination, but, while critically important, animal reintroductions have been a small part of the picture. 'We introduced 451 herbivores, which is hardly anything when you consider that Gorongosa is home to 100,000 large animals,' Marc tells me. 'In South Africa, a lot of projects will do that in a year. But those additions were of tremendous strategic importance because the small herds could not have grown to healthy levels without them, whereas local nuclei of waterbuck and impala were able to rebound with good law enforcement alone.'

This strategy, says Marc, has struck the perfect balance. Over a relatively short time frame, Gorongosa's herbivores rebounded to just below pre-war levels; in turn, the invasive shrub that exploded during their absence has now returned to its pre-war state. By 2018, the total biomass of nine focal species, all herbivores, had reached 95 per cent of pre-war levels (compared with 50 per cent in 2010). Wildlife numbers have grown so well that Gorongosa has begun moving some of its animals to neighbouring protected areas. In total, 2,100 animals (of five species) have helped restock other Mozambican reserves. Gorongosa has also become a model for rewilding in Mozambique and beyond. In the words of one writer, the overhaul of the park is 'a watershed experiment'.[10] A paper Marc co-authored puts it nicely: 'Despite its tragic conservation history, Gorongosa presents an unparalleled opportunity to study the impacts of severe [large mammalian herbivore] declines over large spatio-temporal scales.'[11]

Although hope began bubbling to the surface as early as 1994, few believed the park could return to its former glory. Now that Gorongosa is approaching its pre-war state, the original champions of restoration, many of whom risked their lives to support the project, feel an ongoing pull to the park. Roberto regularly visits, and every year Greg invites Paul up from South Africa. The annual trip that Paul calls a 'soul-feeding catharsis' usually coincides with his birthday, but the pandemic precluded him from travelling to Mozambique when he turned 87 in 2020. Paul has retired – technically – but, like Jane Goodall and David Attenborough, he doesn't let age get between him and the cause. 'You become married to this work,' he says. Coupled with a barrage of negative news headlines and setbacks, conservation's long-term nature makes it near impossible for practitioners to feel like their work is ever done.

It's been 30 years since Mozambique's violent civil war came to an end, and seven since the country became landmine-free.[12] By then, Gorongosa's rebirth was well on its way: several species had been reintroduced while others naturally rebounded in the absence of conflict. Leading scientists from all over the world had begun coming to the park to study its unique wildlife and evaluate how best to protect it, whether by reintroducing herbivores to regulate invasive plants or encouraging local communities to pursue sustainable agriculture in less sensitive portions of Mount Gorongosa.

In recent years, rates of annual forest fragmentation have stabilised (though on average 1 hectare (2.5 acres) vanishes every day)[13]. It is hoped that as additional economic and educational opportunities arise, more locals will practise sustainable land management. The Gorongosa Project has also made a point of creating opportunities for locals to visit Gorongosa, and see and smell it first hand, which is a critical step towards encouraging communities to appreciate more than the park's economic utility. In Africa and beyond, people living in close proximity to protected areas are often disconnected from the wilderness on their doorstep, either because they don't see value in visiting such sites or because they can't afford to.

Tangible connections between nature and local communities are essential, but there is no doubt about the fact that Gorongosa's rewilding required a helping hand and significant economic capital. That hand, however, will not lend help forever. A central tenet of rewilding holds that human involvement must eventually cease or, at least, decrease. Once ecosystems become self-sustaining, conservationists should slacken the reins. Like rewilding itself, Greg's involvement with Gorongosa isn't perpetual. When the time is right, he will fully pass the torch to local entities and exit the stage. 'It's important to remember that Gorongosa belongs to the people,' Greg tells me. 'When I retire, everything will continue just the way it is'.[14]

CHAPTER TWO

Extinct in the Wild

S trapped into the Chevrolet Suburban's hunting seat, I closed my eyes and imagined I was flying. I was a hawk soaring over limestone cliffs, then a duck dancing across the emerald waters of a pond's glassy surface. Like most five-year-olds, I had an imagination so vivid, it entertained me for hours, but I wasn't dreaming when I spotted an African antelope on the Central Texas plains.

'Dad, oryx!' I shouted into the wind.

He couldn't hear me from inside the car.

I turned around and tapped on the windshield, prompting him to brake. As the tank-like SUV slowed to a stop on the gravel road, the sound of crunching rocks swished through my ears.

Dad rolled down his window and poked out his head. 'Where are they?' he asked.

I pointed to the ridge, where I saw a handful of scimitar-horned oryx, their long horns backlit by a rising sun.

Dad winked, 'Good spot, Millie.' I basked in the hard-earned compliment as Dad pressed his boot against the accelerator and sped towards the ridge.

Years later, I learned about the scimitar-horned oryx's plight when the IUCN declared the North African antelope 'extinct in the wild'. Unlike full-blown extinction, species declared extinct in the wild are still represented in captivity. If (or when) captive populations cease to exist, the animal goes the way of the dodo. Even then, there's always the chance that a lone individual or small group of survivors reappears. Discoveries of presumed-dead 'Lazarus species' – like the takahē of

Chapter Three – typically happen by accident, but NGO Re:wild hopes to change that through its 'Search for Lost Species' programme. Re:wild uses rewards to incentivise searches by scientists and members of the public. However, the chance of rediscovering lost animals is remote when it comes to large mammals like oryx or rhinos. For better or worse, large creatures are unlikely to go unnoticed.

On the other hand, captive breeding has helped multiple species deemed extinct in the wild come back from the brink. Animals relegated to captivity aren't lost causes, not yet, particularly when dozens or more unrelated individuals are managed in conservation-minded zoological institutions. 'Since they could go extinct on our watch, restoring *these* species is the ultimate form of rewilding,' says Axel Moehrenschlager, chair of the IUCN specialist group on translocation.

At least five large mammals designated extinct in the wild have returned to their native habitat after captive breeding, or *ex situ* conservation, gave them a critical boost. All of their stories are remarkable: they feature twists and turns and unexpected alliances. In each of them we see the fragility of the natural world as well as its ability to rebound with the help of innovative, forward-thinking conservation. Critics argue that resources should be allocated to species that stand a better chance of survival, but critics don't control purse (or heart) strings, and as one of this chapter's stars points out, passion – not science – is what often pushes conservation over the finish line.

Russell Barnett Aitken was an artist, adventure writer, big game hunter – and conservationist, at least according to a *New York Times* article reporting on his death in 2002.

In his 1969 book, *Great Game Animals of the World*, Aitken devoted a chapter to African antelope or, as he affectionally called them, 'glamor game'.[1] He relished taking down the handsome four-legged beasts; the more challenging the hunt, the better. In 1959, during a jaunt to central Chad, he set a new world record when he shot and killed a massive scimitar-horned oryx, a species he termed the 'lancer of the desert'.[2] Three weeks later, another hunter beat his record and, according to Aitken, at least a dozen larger bulls were 'bagged' in the decade between his trip and the publication of his book 10 years later.[3]

Aitken claimed that all big game hunters are collectors, but as this chapter reveals, not all animal collectors are hunters. For thousands of years, people have captured wild animals for several purposes, from displaying to breeding to domesticating them. Antelope – including oryx – have been kept in captivity since at least 2500 BCE, as evidenced by Egyptian tomb depictions of oryx sporting collars. Zoos have displayed scimitar-horned oryx for hundreds of years and, more recently, private citizens – including my maternal grandparents – acquired the long-horned antelope like they were pieces of living art. For a time, such collections were not unlike the trophies hanging on Aitken's walls: they satisfied the owner's curiosity while showcasing his or her status. But as the scimitar-horned oryx raced towards extinction, captive herds took on greater significance, this time grabbing the attention of bona fide conservationists.

A master of arid environments, the scimitar-horned oryx used to inhabit large portions of the Sahara Desert and Sahelian grasslands. The antelope's range ebbed and flowed with a changing climate, but most experts agree that it began as far west as Mauritania and extended eastwards to Sudan, forming a semi-circle along the border of desert and savanna. At its peak during the early Holocene epoch, a million or more scimitar-horned oryx roamed the African continent.

Since Roman times, habitat loss and hunting have put a dent in scimitar-horned oryx numbers and range. Before trophy hunters came along, people targeted the animal for its meat and hide – the latter was used to make durable shoes, ropes and shields. Increasing drought and desertification further pressured the species, as did resource competition with livestock brought through the antelope's range by nomadic herders. And, like in other chapters, civil unrest played a part in the species' decline.

The earliest known countrywide extinction took place in around 1850 and herds living in the northern portion of the oryx's range likely vanished by the end of the nineteenth century. To the south, scimitar-horned oryx were abundant in the southern Sahelian zone until the 1930s. In some places, the species remained a common sight: a group of 10,000 was spotted in Chad in the 1930s, and 100-strong herds were seen in parts of Niger and Chad as late as the early 1960s.

Herds had shrunk in size, but British biologist John Newby regularly encountered scimitar-horned oryx when he started working in Chad in 1971, leading him to believe the species was stable. He was volunteering with Chad's wildlife department in the Ouadi Rimé-Ouadi Achim Reserve* – the same spot where Aitken hunted 12 years earlier, although it wasn't a protected area until 1969. It later became clear that frequent oryx sightings gave John and the region's nomadic communities a false sense of

---

* The reserve is sometimes called the 'Ouadi Rimé-Ouadi Achim Faunal Reserve'. Elsewhere, 'Wildlife' and 'Game' take the place of 'Faunal'.

security. The species wasn't stable; not in the least. When additional assessments were conducted years later, and global data collated, John and his conservation colleagues realised that he and the oryx in his midst had been living on an island. By 1977, the reserve was home to 95 per cent of Africa's scimitar-horned oryx and contained the world's only viable population.

Between 4,000 and 6,000 individuals remained, but the years that followed spelled disaster as conflict and modern weaponry descended on the region. Chad was already experiencing civil unrest when the Chadian–Libyan conflict commenced in 1978. As fate would have it, the frontline fell smack in the middle of the scimitar's last stronghold. John and his team of local rangers were ill-prepared when fighting spilled into the reserve, bringing with it four-wheel-drive vehicles and automatic weapons. 'We had four ancient guns among a group of 22 rangers,' says John. 'We couldn't do much and soon the reserve was a free-for-all where oryx were slaughtered for their meat.'

A number of rangers joined rebel forces; others abandoned their posts in search of safety. John stayed in the reserve until early 1979, when he travelled to Niger to conduct a comprehensive wildlife assessment. 'I went there to identify new areas that could be conserved to make up for the probable loss of Ouadi Rimé.' John ended up staying in Niger for 11 years as he worked with the World Wildlife Fund (WWF) and the country's wildlife service to establish new protected areas. During his decade-long stint, the English conservationist only spotted a single group of four scimitar-horned oryx. It was the last time he'd see the species in the wild until an unlikely coalition of conservationists teamed up to return it to Chad in 2016.

After nearly 10 years of war, Chad's conflict with Libya ended in 1987. No one knows how many scimitar-horned oryx survived the ordeal: an anonymous 1987 report claimed that 1,000 were counted in Ouadi Rimé following the end of the war, but John sincerely doubts its accuracy. Indeed, he is 'absolutely certain that no large numbers of oryx were around at that time'. In any event, the number of surviving antelope quickly plummeted, and before long extinction became a near certainty. In 1989, a lone male – the last known scimitar in the reserve – was shot and killed. Deep down, John knew the animal was the last wild scimitar-horned oryx on Earth, but others held out hope for survivors living elsewhere. None were found and in 2000 the IUCN officially declared the scimitar-horned oryx extinct in the wild, pointing to habitat loss, competition with livestock, overhunting and civil unrest.

When the last wild scimitar-horned oryx died in 1989, I was an eight-year-old girl living in San Antonio, Texas, thousands of miles away from the ungulate's native range. The word 'conservation' wasn't in my dictionary and I wouldn't travel to Africa for another 20 years, but, fantastical as it sounds, I encountered scimitar-horned oryx on a regular basis. They weren't in a zoo, nor in story books. They were in my midst, in the flesh, living a life that appeared wild, at least to me.

My improbable connection to the African antelope began before I was born, when my maternal grandparents started introducing oryx and other 'exotic' species to their Texas ranch in the early 1970s. Brady Creek Ranch, or Brady as we called it, spanned 8,093 hectares (20,000 acres) of Central Texas. It was a place of shifting landscapes. There were flat pastures trampled by cattle, creeks containing prehistoric (and terrifying) creatures called alligator gars, and rolling hills. One rocky outcropping was decorated with red drawings, many of animals, left by

Lipan Apaches. Another contained caves my cousins claimed were haunted (I was too scared to investigate).

When I was at Brady, I wanted to spend all my time in the so-called 'exotic game pasture', where scimitar-horned oryx and other foreign species grazed on wild grasses. Luckily for me, the house was located inside the pasture. Although there were cattle guards in place to keep large animals out of the yard, you only had to walk along the fence of the sloping lawn or up the winding driveway to encounter a rhea, ostrich or llama. The antelope and zebra stayed further afield; the zebras were particularly shy, making them hard to spot on game drives.

I spent countless hours sitting with my grandparents in the cosy living room at Brady, a space decorated with hunting trophies including the imposing horns of a Cape buffalo that hung over the fireplace. My grandmother would needlepoint cushions and rugs while my grandfather entertained the grandkids with hilarious stories. Many of these memories are razor-sharp – I can still smell the wood-burning fire and taste the peculiar Southern dishes, like turkey hash, consumed at our wooden dining table – but I can't recall talking to my grandparents about how (or why) they acquired animals from far-flung corners of the world. Nor did we discuss big-game hunting, perhaps because they knew that I did not approve.

For me, the animals roaming the exotic game pasture were fixtures at Brady. They were part of its fabric; they belonged. Perhaps this explains why I never investigated their origin. It wasn't until decades later, when I became interested in conservation and learned about the scimitar-horned oryx's extinct-in-the-wild status, that I began thinking about the ones kept at the family ranch. We no longer own them or the property – the land and its animals were sold in 2015 – and the memory-keepers have either passed away, forgotten particulars or never knew the underlying history. Whenever I ask my mom's generation about the exotics, I hear a familiar refrain: my grandfather was one of the first Texans to introduce exotic animals to a private ranch, a practice now relatively common in the Lone Star State. But without historic records, I can't validate the claim, nor can I establish what role, if any, my grandparents played in the scimitar-horned oryx's rewilding story.

I do know that my grandfather spent years on the board of the San Antonio Zoo, a place my family frequented when I was a child. Trailing behind my parents, I'd pause at the small pool where penguins swam graceful circles before dragging my family to the lion and tiger exhibitions. A childhood friend lived so close to the zoo, we could hear lions roaring at dusk when playing in her backyard. Although I cherished weekend mornings at the zoo, I started asking difficult questions when certain aspects of the zoo experience began to trouble me. I'd ask my parents why some enclosures were smaller than others; and I wanted to know why several featured natural elements, like trees, while others were constructed entirely of concrete.

Ethical concerns notwithstanding, I don't remember asking my parents about the origins of zoo animals. Just as I glossed over how foreign species ended up on my family's ranch, I never gave much thought to the wild ancestors of modern zoo inhabitants. In my case, the connections finally became evident during a 2015 trip to Botswana. I was in the middle of the Kalahari Desert visiting the otherworldly Makgadikgadi Pan when I met 'Cobra', a member of a local Khoisan tribe in his 70s or 80s. Cobra had a penchant for digging scorpions out of their burrows, but he didn't stop at removing them with his bare hands. He wanted me to see the intricacies of their bodies. They were covered in dirt, so he popped them in his mouth to clean them off – all while avoiding getting stung. In light of Cobra's nickname, I was unsurprised when he told me he was adept at capturing a range of animals, including dangerous ones. This, he said, made him extremely valuable to representatives of zoological institutions that visited the region decades earlier to stock their collections. I remember feeling shocked, even unbelieving. Was this really how zoos came to possess animals from around the world?

I was desperate to ask Cobra about his live capture techniques and the relationships he built with global zoos, but our time together was limited, so I left Botswana with more questions than answers. The following year, my curiosity was finally sated when a friend gave me a copy of David Attenborough's memoir, *Adventures of a Young Naturalist*. Attenborough recounts his experience helping the Zoological Society of London acquire animals for its zoological collection while filming *Zoo Quest*, a popular BBC wildlife documentary programme that launched his career as a presenter. His memoir opens with the following sentence, 'These days zoos don't send out animal-collectors on quests to bring 'em back alive. And quite right too.'[4]

Animal historian Nigel Rothfels has spent decades researching and writing about modern-day zoos. He tells me that once initial collections were underway, their growth was often exponential. For instance, in 1909, a mere 10 years after it was founded, the New York Zoological Park (now the Bronx Zoo) housed 743 mammals, 2,816 birds and 1,969 reptiles. The zoo's antelope collection wasn't noteworthy until the 1903 addition of the Antelope House. That summer, a bulletin published by the zoo said it was 'seeking far and wide for the species of hoofed and horned animals'. Some would arrive from Europe; others were 'on the way from the interior of Africa'. German animal dealer Carl Hagenbeck was the zoo's go-to supplier. Hagenbeck sold a wide variety of species to zoos around the world, in some instances stocking an entire facility with one order, but he struggled when it came to procuring antelope. In a letter to the New York Zoological Society's then-director, William Hornaday, he wrote, 'I must tell you, it is the hardest job to collect antelopes.'[5]

In the end, Hagenbeck sold numerous hoofed animals to the New York institution, and when the Antelope House opened to the public in 1903 it contained 13 species. Scimitar-horned oryx weren't among them, perhaps to the species' detriment since Hornaday was ahead of his time when it came to the role zoos can play in wildlife conservation. In 1910 – 70 years before zoos banded together to create what are called Species Survival Plans – he wrote, 'It is indeed time for the men of today who care for the interests of the men of tomorrow, to be up and doing in the forming of collections that a hundred years hence will justly and adequately represent the vanishing wild life of the world.'[6] (Hornaday, it must be said, was not progressive on all fronts. In 1906, he put a Congolese man named Ota Benga on display in the zoo's Ape House. Swift rebukes followed, and Benga left the zoo soon after, but Hornaday attempted to justify his decision, claiming that he was attempting to offer the public an "ethnological exhibit". Benga, sadly, killed himself 10 years later.)

By 1910, scimitar-horned oryx had already vanished from the northern fringes of the Sahara, but robust populations in several native range states meant they probably weren't on Hornaday's mind when he wrote about the urgent need for zoo-assisted *ex situ* conservation. Nor were zoos motivated by conservation when they started adding scimitar-horned oryx to their collections half a century later – even though the animal's status was more precarious by then. If John Newby thought the species was relatively stable in its southern range in the 1970s, zoos in the United States, Britain and Europe – the first regions to acquire the antelope in the 1960s – couldn't have foreseen the animal's fate (or the part they'd play in reversing it).

Experts don't know when the first scimitar-horned oryx appeared on zoo grounds, but historic records tell us that several American institutions possessed scimitar-horned oryx in the 1940s and 50s. Many of these records were retroactively added to the scimitar-horned oryx studbook – a breeding registry that details when animals are born or die, where they originated and when they are transferred between facilities. None of them, however, contain origin information. On the other hand, dozens of entries from the 1960s reference wild-caught scimitar-horned oryx that originated in Chad.

These records are critical because they reveal information about the ancestors – or founders – of today's captive herds. Such records help answer questions like: where did the founders come from? Did the scimitar-horned oryx of today descend from 10 closely related individuals, or was the founder population sizeable and genetically diverse? Answering these questions isn't a historic exercise. Since the future of the species depends on rewilding captive-bred animals, scientists need to know what they're working with from a genetic standpoint.

Because of the studbook, we know that most of the scimitar-horned oryx alive today descended from 40 to 50 wild oryx captured in Chad in the 1960s and sent to zoological institutions in North America, Europe and Japan. A Dutch animal trader orchestrated two capture events – one in 1963 and another in 1967. The second retrieval was more substantial than the first: it involved

44 animals, more than half of which went to the United States (the US received 26, 16 stayed in Europe and Japan received a pair). Of the batch sent to the US, five made their way to the San Antonio Zoo.

They settled into their new lives with relative ease; indeed, calves were born the following year. Most members of the first generation born in San Antonio were sent to other zoos, but I eventually discovered that three of them were sold to my grandfather. He acquired two in 1971 and a third in 1975. Two males and one female, all born to wild-caught parents, launched the herd I spent many weekends admiring. According to the San Antonio Zoo's archivist, my grandfather was one of only three Texas ranchers to acquire scimitar-horned oryx from the zoo in the early 1970s. He was, it seems, a captive oryx pioneer after all.

Curiously, these private transactions do not appear in the studbook. I only know about them because San Antonio Zoo archivists have kept meticulous records, which begs the question: if the records exist, why not add them to the studbook? The current archivist isn't sure, but I suspect that it boils down to the fact that most institutions only kept records of animals they considered relevant in the long run: those that could be bought, sold and bred. The oryx sold to private ranchers were, I imagine, considered ancillary. Their value wouldn't become clear until many years later.

In the meantime, captive herds living in zoos and on private lands grew exponentially. 'Scimitar oryx breed like rabbits,' says John, 'and are very easy to look after, so an enormous paradox arose where you had a species thriving in captivity that went extinct in the wild.'

The 50-odd scimitar-horned oryx sent to zoos in the 1960s bred so well that hundreds, and later thousands, of their descendants populated zoos and ranches. No one knows exactly how many oryx currently live on private lands in the United States and Europe, but experts believe that upwards of 12,000 scimitar-horned oryx are found on Texas ranches alone. Some are hunted – Texas has a massive exotic hunting industry – while others are part of safari-esque attractions. Herds like the one that lived on our ranch are kept solely for their owners' viewing pleasure[*].

In contrast to private owners, zoos have kept records for decades. The studbook tells us that, in total, 573 global zoological institutions have possessed scimitar-horned oryx; in 2021, around 1,750 individuals were displayed in 194 international zoos. The oryx living in zoos are often managed with specific goals in mind. Beginning in the 1980s, accredited zoos bred (and traded) threatened species under the auspices of Species Survival Plans (SSPs), keeping what are known as 'insurance populations'. If a tragedy befalls the species – a plague, say, or civil war – captive groups can be used to repopulate the wild. By the 1980s, zoos and conservation organisations recognised that captive animals had the potential to save species from the greatest threat of all: extinction.

Scientists believe that the first conservation-minded species reintroduction occurred in 1907 when Hornaday teamed up with the American Bison Society – an entity he

---

[*] My family members and I can't recall how many oryx lived on the ranch at any one time, but my parents estimate that Brady's herd typically contained approximately 15 individuals. The ranch manager evidently traded individuals on occasion to prevent inbreeding, but the population was not otherwise managed (as it would be in a zoological institution). When the ranch was sold, its hooved animals were, too. Given the large number of scimitar-horned oryx residing on Texas ranches, it would have been easy for the ranch manager to find new homes for the oryx, zebras and other animals held in the exotic game pasture. I, unfortunately, had no say in any decisions pertaining to the ranch since I did not own the land or any of its assets.

and Theodore Roosevelt helped to create – to ship 15 American bison from the Bronx Zoo to a reserve in Oklahoma.[7] The first *high-profile* reintroduction wouldn't take place until the 1980s, however. As fate would have it, the initiative involved a close relative of this chapter's protagonist: the Arabian oryx.

The Arabian oryx teetered on the brink of extinction well before its North African cousin: the last wild animal was killed in 1972 – 17 years before the scimitar-horned oryx went extinct. Conservationists had seen extinction coming for nearly 20 years and intervened just in time, retrieving several wild animals from modern-day Yemen in 1962 under a programme called Operation Oryx. Unlike the 1960s captures of scimitar-horned oryx in Chad – which were commercially motivated – Operation Oryx had captive breeding and conservation in mind from the get-go.

Operation Oryx was led by the Fauna Preservation Society (now Fauna & Flora International) (FFI) in partnership with the WWF, but its success hinged on local know-how. Operation Oryx might have failed had it not been for a local tracker who possessed what FFI called 'almost supernatural tracking ability'.[8] The high-profile retrieval generated praise from the very top, with Prince Philip commending FFI and its partners in a 1962 telegram.

Very glad to hear of your remarkable success in capturing three Arabian oryx so that they may breed in captivity and thereby save the species from extinction. The cooperation of the Fauna Preservation Society and the World Wildlife Fund in this Noah's Ark operation is a splendid precedent for future efforts to save the world's endangered species.

Soon after, a small number of captive Arabian oryx –
including several owned by the leaders of Kuwait and Saudi
Arabia – joined three animals retrieved from Yemen at the
Phoenix Zoo in Arizona. Together they formed a 'world
herd'. It grew steadily, and in 1978 the species returned
to its native habitat for the first time when the Phoenix
Zoo sent four animals to Jordan. They were released into
a fenced reserve but never allowed into the open desert
due to limited space and security concerns. A 'true
reintroduction' commenced four years later when a cohesive
group of 10 animals was released into an Omani protected
area later named the Arabian Oryx Sanctuary. Additional
reintroductions have since taken place in Israel, Saudi
Arabia and the UAE, and there are currently between 850
and 1,000 Arabian oryx running wild in the Middle East.
Remarkably, the IUCN later reclassified the species as
'vulnerable', three levels below 'extinct in the wild'.

However, the decades-long project has experienced its
share of setbacks. Poachers came first, decimating the
population in the mid and late 1990s after it grew to more
than 400 individuals (perversely, the herd's success caused
the poaching spike). Then came a massive blow from
Oman's government. In 2007, UNESCO pulled the plug
on the Arabian Oryx Sanctuary's status – the first and only
time the body has done so – after Oman decided to decrease
the reserve's size by 90 per cent. It's unclear how many
oryx live there today, but even if the Omani population
doesn't stand the test of time, reintroductions in the Middle
East have paved the way for rewilding efforts elsewhere.

The success of Arabian oryx reintroductions
undoubtedly influenced the first trial release of scimitar-
horned oryx in 1985 – four years before the last wild
scimitar-horned oryx was killed in Chad, and 15 years
before the IUCN declared the species extinct in the wild.[9]

Ten scimitar-horned oryx from British zoos were sent to Tunisia, but because the herd lived in a fenced, albeit large, enclosure inside a national park, the release wasn't considered a reintroduction, mirroring Jordan's Arabian oryx experience. It was nevertheless significant, both for Tunisia – a country that hadn't possessed scimitar-horned oryx for 80 years – and the long-term initiative that ultimately returned the species to the wild. Let's call it Operation Scimitar-Horned Oryx.

In the decades that followed the initial release in Tunisia, additional 'soft releases' were carried out in three other Tunisian protected areas, a national park in Morocco and a nature reserve in Senegal, but none of the scimitar-horned oryx returned to these sites ever left fenced enclosures and all six populations continue to be actively managed. For these reasons, the releases don't qualify as 'true reintroductions', at least not in the eyes of reintroduction expert Mark Stanley Price of Oxford University (Mark previously managed the Arabian oryx project in Oman and founded the IUCN Reintroduction Specialist Group). An entire book could be written about why the distinction matters, but the simplest explanation cuts to the core of rewilding. Rewilding seeks to restore *natural* orders, processes and rhythms. Most rewilders agree that people can *facilitate* nature's recovery, but perpetual management compromises the very wildness they are trying to achieve.

John wasn't involved in these trial releases, nor was he in North Africa to witness them. In 1991, he left Niger for Switzerland to take a job at the WWF's headquarters. John's passion for North Africa and desert animals never waned, however, and his longstanding dream of returning scimitar-horned oryx to an unfenced wilderness area began to crystallise when he visited Abu Dhabi in 1998 to explore opening a WWF office there. He had heard about several private collections of scimitar-horned oryx in the area, the largest of which belonged to Sheikh Zayed Al Nahyan – the country's founder and long-time ruler – and kept on his private island, Ṣ īr Banī Yās, which is located 160km (100 miles) from the capital city. A man who worked for one of the Sheikh's sons invited John to visit the island. During a day-long private tour, John saw hundreds of scimitar-horned oryx. On a trip several years later, he counted 1,400 of them.

It wasn't a discovery per se: many knew that Sheikh Zayed's menagerie existed, and the site had welcomed weekend visitors for years. The island's connection to conservation began in 1977 when Sheikh Zayed banned hunting on its shores and launched a captive breeding programme of sorts. It started with Arabian oryx and expanded to include not just scimitar-horned oryx but also giraffes, ostriches, caracals, and 20-odd other native and foreign species. Seeing the island's large scimitar-horned oryx herd nevertheless set off a lightbulb in John's mind.

Like the oryx living on Texas ranches, Sheikh Zayed's population hadn't been managed for genetic diversity purposes. They weren't part of an SSP, and their origin was murky (the herd's founders probably came from Chad and Sudan, but no one knows for sure since the animals were gifted to Sheikh Zayed). Even so, because they and their ancestors lived (and bred) apart from the rest of the world's scimitar-horned oryx population, the Emirati oryx offered genetic diversity not found elsewhere.

Their relative proximity to the species' native range in North Africa was also appealing, as was the herd's habituation to arid environments. Most importantly, their billionaire owner cared about the species and wildlife conservation more broadly. Sheikh Zayed was already in his 80s when John first visited his wild island, and the charitable leader passed away a few years later, in 2004. Sheikh Zayed's legacy is vast – and varied. The founding father of the UAE governed the country for almost four decades. He launched and supported numerous charitable causes, but his establishment of the Environment Agency of Abu Dhabi, or EAD, in 1996 stands out. The organisation spearheads multiple initiatives, including the captive breeding and rewilding of endangered animals, a programme that grew out of Sheikh Zayed's early work supporting Arabian oryx.

In 2002, John decided to move to Abu Dhabi to work for EAD. He'd been with the WWF for 28 years and was looking for a new challenge. More importantly, he missed working in desert ecosystems. 'At the time, none of the big NGOs were focused on desert species,' he says. Although John enjoyed his work in the Gulf, he didn't stay at EAD for long – in 2004, he struck out on his own to launch the Sahara Conservation Fund (SCF) – but the relationships he cultivated at the agency were invaluable.

John recognised that more than any global institution supporting scimitar-horned oryx conservation, EAD had the resources required to make rewilding a reality. American conservationist Justin Chuven felt the same way. Like John, Justin spent decades fantasising about the scimitar-horned oryx's return to the wild, and nine years as the San Diego Zoo's lead mammal keeper gave him intimate knowledge of scimitar-horned oryx and captive breeding practices. So, when EAD invited Justin to lead its growing *ex situ* conservation programme in 2011, he jumped at the chance.

Together, John, Justin and a colleague at the Smithsonian Conservation Biology Institute started plotting the creation of a scimitar-horned oryx world herd that would, like its Arabian oryx predecessor, foster genetic diversity among animals destined for the wild. The first members of the herd came from the late Sheikh Zayed's private collection, which EAD took over following his death. Soon after, the agency teamed up with the Royal Zoological Society of Scotland to sequence the herd's DNA, finding that the animals descended from a small number of maternal lineages – an unsurprising discovery given that Sheikh Zayed's herd likely began with a handful of animals.

It was then that the value of collections outside the Middle East became clear. Justin would need to gather a truly global world herd to give the scimitar-horned oryx the best chance of survival in the wild given the limited number of oryx that founded today's population. Fortunately, many conservation-minded zoological institutions were actively looking for opportunities to help rewild the species, and Justin already knew many of the key players – among them Tania Gilbert and Adam Eyres. Tania is a conservation biologist with the UK's Marwell Zoo. Marwell holds the global scimitar-horned oryx studbook and Tania manages it on Marwell's (and the world's) behalf. Adam is from my neck of the woods. He's the director of animal care for the Fossil Rim Wildlife Center – a private ranch-turned-zoo in Texas. Although Justin primarily looked to foreign herds when sourcing new oryx, he also investigated several private collections in the Middle East and was pleasantly surprised to discover scimitar-horned oryx in Dubai whose maternal lineages were different from those present in Sheikh Zayed's collection. Justin scooped them up while exploring strategic global acquisitions.

Building a robust world herd was a Herculean task, and it was just one piece of the puzzle. First and foremost, funding had to be secured – and lots of it. Moving animals between regions is incredibly expensive, explains Adam, who has managed numerous oryx shipments from Fossil Rim. In total, the Texas facility has sent 18 scimitar-horned oryx to

Abu Dhabi over the years. By Adam's estimate, it costs around $1 million to send 25 oryx from Texas to Abu Dhabi.

A breeding centre had to be built, too. Justin estimates that EAD spent around US$20 million building the 230-hectare (568-acre) Deleika Wildlife Conservation Centre in Abu Dhabi. It became operational in 2010 and welcomed its first two batches of foreign-born scimitar-horned oryx in 2014 and 2015. Six American facilities (including several private ranches) sent 42 animals, and British and European zoos sent an additional 14 animals. At Deleika, scimitar-horned oryx were divided into enclosures based on gender and place of origin so that males from the UAE would breed with females from abroad and vice versa. Every animal underwent health checks to ensure they were disease-free, and during the process blood samples were collected for further DNA analysis. Justin played matchmaker to maximise genetic diversity in future generations.

Meanwhile, genetic experts like Rob Ogden (who then worked for the Royal Zoological Society of Scotland) helped the team calculate the number of scimitar-horned oryx needed to repopulate the wild to ensure the species' long-term success. There is no magic number, says Rob, who underscores that a range of factors impact 'population viability analysis', but 500 is a common benchmark. EAD and SCF rallied behind it, declaring the goal of seeing a 'viable, free-ranging and self-sustaining population of at least 500 animals'. In reality, the goal is more nuanced.

Although Operation Scimitar-Horned Oryx would be delighted to see 500 scimitar-horned oryx roaming the wild – a milestone already within reach – Rob emphasises that a healthy, self-sustaining population will require more than 500 animals. That's because an important distinction exists between a 'census population size' (the total tally of animals) and an 'effective population size', which, in Rob's words, only includes 'breeding adults that contribute equally to the next generation'. Rob estimates that the scimitar's census size is around 10 per cent of the effective size – so the reserve's population would ideally number around 5,000, not 500, oryx for it to be genetically sustainable. If the percentage seems low to you, bear in mind that the scimitar-horned oryx is a harem-forming species wherein a small number of males mate with a large number of females. As a result, the number of breeding adults, on its own, is not an instructive figure. In any event, the jump from 500 to 5,000 isn't massive, says Rob, given the species' proclivity for prolific breeding. However, John would like to conduct a fresh population viability analysis that plugs in new, real-life information gathered in the field. He's keenly aware of the fact that population modelling tends to produce optimistic, best-case-scenario predictions whereas life in the wild is, as we know, inherently unpredictable.

At the same time, innovations in the field of molecular genetics mean that Rob and experts like him can better understand genetic diversity across the wider population, and DNA sequencing can fill in gaps for animals whose records are scant or non-existent. Texas ranch populations could become even more important in the future if testing identifies maternal lineages distinct from those already represented in the world herd. 'Whole genome sequencing has created potential for a lot more accurate conservation of diversity within breeding programmes,' says Rob. 'But this technology is brand new. We're on the cusp.'

While EAD focused on creating a healthy herd, John drew from his work in North Africa to answer a paramount question: where should reintroductions take place? Where could the species survive – and, more importantly, thrive? He and others tasked with considering potential reintroduction sites were hindered by the fact that much had changed since John left Niger for Switzerland in 1991. Desertification had increased, as had development; and according to John, conservation infrastructure was virtually non-existent in countries like Chad and Niger. 'We had to do serious due diligence,' says John, who collaborated with the Zoological Society of London to initiate the Pan Sahara Wildlife Survey. The two-year-long survey utilised geographic information system (GIS) mapping to evaluate eight sites in three countries: Chad, Niger and Tunisia. It also gave conservationists the opportunity to build relationships with national authorities while helping them understand the status of the region's biodiversity. In the end, survey data demonstrated what John suspected all along: Chad's Ouadi Rimé-Ouadi Achim Reserve was the best site for the scimitar-horned oryx's return to the wild.

Its first asset was size: at 7.8 million hectares (19.3 million acres), the reserve is as big as Scotland, making it a prime rewilding destination for a species that requires large stretches of land to survive. Ouadi Rimé also contains suitable habitat – and enough of it – for herds to grow; though challenges, like resource competition with livestock, could increase over time. Finally, unlike some of the other countries in the species' historic range, Chad seemed politically stable.* John called the site's selection a 'no-brainer'.

---

* More recently, political instability has shaken Chad. In April 2021, the country's decades-long leader, President Idriss Déby Itno, was killed while visiting troops battling a rebel group that penetrated northern Chad from Libya. Addressing the conflict in September 2021, John provided assurances that Ouadi Rimé has not been impacted by escalating tensions; and in early 2022, Chad's transitional government released a number of rebels from prison while inviting rebel groups to join peace talks scheduled for February the same year.

It helped that he had considerable experience in the
region. Contacts, too. 'The kids I played soccer with
decades ago are now in the upper echelons of Chadian
society,' explains John. 'These enabling factors are critical,
as are cultural and historic linkages.' Because Ouadi Rimé
was the scimitar-horned oryx's last stronghold, returning
the species to the reserve meant bringing things full circle.
Elder members of local nomadic groups still remember
living alongside the species and welcomed its return. John
spoke to numerous communities in advance of
reintroductions, but one particular visit stands out.

He was in the field, sitting beneath the shade of an Acacia
tree, as he spoke to locals about the proposed reintroduction
project. The Chadian men and women who'd gathered
around him expressed their enthusiasm for the oryx's
comeback, but as the meeting came to a close, one local
turned to John and said, 'For God's sake, please don't
introduce more hyenas!'

Unlike predator reintroductions that spark safety
concerns, antelope rewilding has the benefit of posing few
threats to people, and local buy-in is critical. It suggests
that local communities will act as environmental stewards,
whether by refraining from hunting or allowing antelope
to graze near their livestock.

Chad's government was equally open to the return of an
iconic species. When John discussed the initiative with the
late President Idriss Déby Itno, the biologist emphasised
that scimitar-horned oryx are part of Chad's heritage. John
reminded President Itno that the species was taken from
the African nation, making its reintroduction a matter of
cultural significance *and* justice. The Chadian government
embraced the project but was unable to pay for it (according
to the IMF and World Bank, Chad's GDP is among the 15
lowest in the world). Nor were organisations like the SCF

able to finance the reintroduction. Thanks to its billionaire founder, EAD had ample financial resources on top of thousands of scimitar-horned oryx and an *ex situ* conservation manager keen to bring the project to fruition. Despite this, the agency's financial backing was never guaranteed.

Why would an organisation based in Abu Dhabi (and focused on the Emirates) spend millions of dollars on an African rewilding project? A born sceptic, I grappled with this question for months, posing versions of it to everyone I interviewed. Time and again, they pointed me toward Sheikh Zayed's passion for conservation. Oryx, I'm told, were of particular interest to the man whose conservation work earned him a WWF Gold Panda Award in 1997. Together, his role in Operation Oryx and his long-time scimitar-horned oryx breeding programme suggest that he would have eagerly supported the species' return to the wild, but he died before the project began to take shape, leaving EAD the steward of his environmental legacy.

Justin and his boss at the time, Fred Launay, who now helms wild cat NGO Panthera, brought the idea to EAD higher-ups. They highlighted Sheikh Zayed's values while emphasising the unique position the UAE found itself in given the country's large, genetically important scimitar-horned oryx herd. In John's opinion, Sheikh Zayed's legacy can't be overstated, but he also believes that EAD stood to benefit from leading the project. By launching it, the agency elevated its reputation as a serious conservation entity. During one of our calls, after John and I bonded over our shared belief that successful conservation requires compelling storytelling, he repeated a piece of advice he had received years earlier, 'To get Emiratis to support a project, you have to tell them a story they can dream about, because they already have everything money can buy.' The story, in this case, was about honouring a powerful and benevolent man's legacy while helping a critically endangered species. John appreciates the power of storytelling and believes that conservation is a social science *implemented* via technical science. 'Passion is more contagious than science,' he adds. 'It's what opens the door. Scientists hate that, but Goddamnit, it works.'

Justin must have told his bosses one hell of a story, because they accepted his proposal without hesitation. And just like that, a long-held dream became a reality. The project was poised to move forward – at record speed.

On 16 March 2016, scimitar-horned oryx stepped foot on Chadian soil for the first time in 27 years. The 3,726-km (2,315-mile) journey from the Deleika Wildlife Conservation Centre in Abu Dhabi to the Ouadi Rimé-Ouadi Achim Reserve lasted 31 hours – a taxing trip for a caged animal, but a blink of an eye compared with the 14 years conservationists spent planning the reintroduction. The animals weren't wild, though – not yet. They needed to acclimatise to conditions in Chad, so they spent six months under observation while living in a large fenced enclosure inside the reserve.

Every scimitar brought to Chad had spent at least six months in the UAE, so even the ones that began their lives in rainy England had experience with desert climates, but few were accustomed to grazing on wild vegetation since in their former homes pelleted food was the norm. Water was topped up during their acclimation period, which occurred during one of the region's many dry months, and the dangers facing wild antelope – from predators to poachers – were non-existent during their time behind the fence.

So, when the gates of their temporary enclosure opened on 16 August, Chad's first wild scimitar-horned oryx in 20 years faced uncertain futures. They were free but at risk; wild but apprehensive. Biologically, the antelope is perfectly adapted to life in the desert, able to survive during months of drought and dizzyingly high temperatures, but conservation managers had to ask themselves this: when it comes to captive-born animals whose ancestors have been in captivity for generations, are biological instincts enough? The first encouraging sign came when the herd moved into the wider reserve hours after the enclosure gates opened. Even more promising was the fact that the released oryx didn't turn to supplementary water and food left out by project managers. It helped that Chad's new oryx were released during the resource-rich rainy season, when grasses are abundant. The decision was strategic but imperfect: pastoralists and their livestock flock to the reserve during the wet season, which caused the team to worry that conflict might erupt (fortunately, none has, though livestock often dominate the best patches of the reserve, relegating wild animals to sub-par habitat).

Although the SCF is the project's in-country implementer, the NGO contracts some of its work out to other organisations, as is common in conservation. Tim Wacher of the Zoological Society of London leads the local team responsible for monitoring the reintroduced oryx on a day-to-day basis, working in tandem with Smithsonian's Katherine Mertes, who tracks the oryx from afar using GPS tracking collars. Remarkably, nearly every released animal wears one, which helps Tim and his trackers ensure the oryx are safe while giving Katherine the chance to study their habits and preferences. Since the species hasn't been in the wild since the 1970s, many knowledge gaps need filling. Better understanding of the species (and how reintroduced antelope behave) will bolster scimitar-horned oryx conservation while building a critical knowledge base from which future projects can draw.

Within 48 hours, the rewilded herd travelled between 30 and 40km (19 and 25 miles). Satellite collar data tells Katherine that the newly rewilded oryx utilise a total area of 3,000km² (1,158 square miles); on average, the herd, whose social structures ebb and flow, moves 15km (9 miles) per day, and most members stay within 50–60km (31–37 miles) of the release site, although several have travelled much further. One, a female known as 'Red 91', travelled to the far north of the reserve and back south within a two-month period. In the context of reintroduced scimitar-horned oryx, her journey was anomalous, says Katherine, but data gathered by John in the 1970s suggests that a long north–south movement during the wet season is 'highly appropriate'. Unlike the scimitar-horned oryx translocated to Morocco, Senegal and Tunisia, the oryx reclaiming Chad are free to go wherever the wind blows them (or, more aptly, wherever the rain takes them), making theirs a genuine reintroduction.

Since their release in August 2016, the first transplants have successfully navigated multiple eight-month-long dry seasons, travelling long distances to locate food and water when necessary. Six months later, in early 2017, conservationists sent a further 14 scimitar-horned oryx bounding into the Chadian reserve – this time during a dry spell. American zookeeper Dan Beetem was there. Having worked with the species for decades, Dan was eager to witness the release, but he was even more ecstatic about tracking the first herd – the one set free the prior August. In it were three scimitar-horned oryx born and raised at the zoological park he manages (The Wilds) and Dan was determined to observe them walking free.

'As soon as I arrived at base camp, I sat down with Tim Wacher to identify the collar numbers for the oryx from The Wilds,' says Dan. He easily located one of the three, a male who was sniffing around the pre-release pens like a lonely soul looking for love at the pub. The other two required tracking. Tim took Dan out to look for them, using the standard combination of satellite tracking data and VHF tracking, and eventually came upon a small herd containing the two animals. 'Once we sighted a group of animals, I could use my binoculars to read an individual's tracking collar,' explains Dan. When he finally spotted the Ohio-born oryx, the immensity of the project hit Dan like a ton of bricks. 'Seeing them standing there in a large group, hours from the nearest paved road, made it real for me.' Back home, Dan had numerous stories to share with colleagues and friends, but he can't recount his emotional encounter without getting choked up – even now.

In August 2017, a much larger release occurred when 54 scimitar-horned oryx joined Ouadi Rimé's growing herd. Already, 22 calves had been born in the wild, bringing the reserve's population size to just under 90. Animals die, of course, but only one perished because of illegal hunting, and the man who killed the oryx knew nothing about the rewilding project since he had just moved to the area from several hours away. After shooting the oryx, he noticed its collar and immediately regretted his action (locals may not know a collar's purpose, but they recognise that they convey ownership). He accepted responsibility for his mistake and spent several months in jail.

Guided by the goal of seeing 500 wild scimitar-horned oryx in the reserve, Justin and his team send new individuals to Chad from Abu Dhabi's world herd twice a year, their timing dictated by logistics and weather conditions. Covid delayed several shipments, but as of early 2021, nine batches had been sent to Chad. Inspired by the project's success, EAD and SCF wasted no time applying their rewilding template to the conservation of several additional desert species.

In 2020, the two organisations turned their attention to addax, a fellow North African antelope that has fared almost as poorly as the scimitar-horned oryx. The critically endangered ungulate contains fewer than 100 wild survivors and was wiped out of Ouadi Rimé at the same time as scimitar-horned oryx, so a pilot reintroduction of 15 in early 2020 was monumental. A second batch containing 25 addax arrived months later and 12 healthy calves were born by the end of the year. At the time of writing, approximately 50 live in Ouadi Rimé alongside just under 400 scimitar-horned oryx.

Chad's scimitar population continues to expand as newly mature oryx mate for the first time. In 2020, several *wild-born* females gave birth to calves, marking a new generation of parents – and calves. 'It's a sure sign that the population is well on its way to becoming viable and self-sustaining,' explains Justin, who predicts the herd will grow to 500 individuals within the next few years. He and his project partners won't rest, however, until the species is downgraded from 'extinct in the wild'. Even then, the beige-and-white oryx will have a long way to go before becoming a common sight again, but successful rewilding in Chad will inspire similar efforts elsewhere while generating interest in the 'lancer of the desert' and the unique ecosystem it inhabits.

The Land of Birds

Things would have gone differently if rats hadn't travelled on the vessels that brought Polynesians, and later Europeans, to New Zealand. The land ruled by birds was virtually mammal-free when Polynesian men and women began settling its virgin shores during the late thirteenth or early fourteenth centuries. The country's only native terrestrial mammals are small bats that spend much of their time on the ground, a common Kiwi adaptation. Without mammalian predators chasing them into the skies, many of New Zealand's winged creatures ditched their powers of flight and evolved other, more appropriate traits. Cruelly, the very adaptations that allowed them to thrive in an avian paradise later became their greatest weakness, their Achilles heel.

In the absence of written records, we can only speculate about what New Zealand's initial settlers were thinking when they discovered their new homeland, but they must have been exhausted after making a 3,000-km (1,864-mile) journey – via canoe – from East Polynesia. Perhaps they were scared, too, having considered the possibility that they'd struck upon a hostile land, a place lurking with dangerous animals or spear-wielding people inclined to attack newcomers. Unfamiliar sights and smells would have abounded: how would settlers know which plants were safe to eat and which were poisonous? Before long, they would encounter some of the land's curious creatures. The giant moa, a massive wingless bird larger than an ostrich, must have been a shocking (if not terrifying) sight. Adapting to a novel environment couldn't have been easy, but when it came to natural resources, the forefathers of the Māori people hit the jackpot. New Zealand was packed with wild protein – and it was all theirs.

There were fish aplenty, fur seals and birds. So many birds. There were songbirds and parrots, penguins and owls. Describing the 'rich general and specific variety of birds would require a ponderous catalogue', wrote one historian.[1] Of the dizzyingly diverse avifauna, moas occupied a special place in the ecosystem. When Polynesian canoes landed on New Zealand's shores, nine species – all flightless (and wingless) – lived on the North and South Islands. They were the country's dominant herbivore, its elephant, and scientists estimate that between 500,000 and 2.5 million were present at the time of Polynesian settlement.[2] Only one predator, the stupendous Haast's eagle, whose wingspan stretched to 3m (10ft), hunted the giant moa. The largest moa weighed 227kg (500lb); by contrast, African ostriches can only reach 113kg (250lb).

Unfortunately, long legs and sturdy bodies didn't protect moas against their new foes – even though early Māori communities didn't possess bows and arrows, slingshots or other 'projectile weapons'.[3] How they took down these weighty birds without such tools remains open to debate, but dogs brought from Polynesia likely helped and the moa's inability to fly definitely hurt. In reality, moas were probably easy targets: they had no experience with ground-dwelling predators so wouldn't have feared humans. In the blink of an eye, all nine species were wiped out. Moas vanished within a shockingly short period of time – as little as 80 years and no more than 160. Theirs has been called the planet's fastest-known megafauna extinction, 'the best instance of overkill, of blitzkrieg, on the record'.[4]

Before long, the moa's lone natural predator vanished, too, as a range of birds, many flightless, tumbled towards extinction.

Throughout history, mass extinctions have occurred every time humans colonised new parts of the world, but New Zealand's story stands apart. To understand the country's biodiversity – and why it was uniquely vulnerable to the colonisation that began seven centuries ago – we have to go back to the very beginning, when the landmass that became New Zealand split off from supercontinent Gondwana 80–85 million years ago. When New Zealand became its own landmass, marsupials and mammals had not yet evolved, so the roles they normally play (and ecological niches they normally fill) went to birds, insects and reptiles. Bats – the country's only native mammal – arrived later, flying from Australia when the two countries were closer together. Some of the birds described in this chapter, kākāpō among them, did the same. Others already present continued their evolutionary journey in a place that's been called 'an evolutionary laboratory'.

One of the last places on the planet to be colonised by people, the 1,610-km (1,000-mile) long island chain known as New Zealand (*Aotearoa* in Māori) was on its own for a very long time and is 'distinctly isolated' from other landmasses.[5] Their isolated nature means that islands typically contain a high number of endemic species (those found nowhere else) and, in the absence of mammalian predators, New Zealand's endemics became highly specialised, with flightlessness a common attribute. Flight requires considerable energy. If you don't need it, why spend it?

In the face of human colonisation, birds living on islands are often at high risk of extinction, with flightless birds typically suffering the highest rates of extinction. They were often the largest terrestrial animal living on archipelagos like New Zealand and Madagascar, making them appealing targets for hungry hunters. But the massive moas that vanished after Polynesian settlement began weren't the only birds lost during the country's first phase of colonisation. In total, 32 bird species went extinct between the Polynesians' arrival and the European colonisation that began five centuries later. The Māori people acknowledged the damage they caused and began to see extinction as a metaphor for their own perilous position following European colonisation (one of their traditional sayings, or *whakataukī* , means, 'The people will disappear like the moa'). Unfortunately, European settlers accelerated environmental declines through habitat change and destruction and the introduction of additional pests, causing another nine bird species to go extinct. To put this into perspective, human impacts (including the introduction of pests) caused 40 per cent of New Zealand's terrestrial birds to go extinct, and 80 per cent of the bird species remaining today are threatened with extinction (as are 100 per cent of New Zealand's frogs and 88 per cent of the country's lizards).[6]

In addition to directly harming native wildlife, invasive species create negative cascading effects on the environments they inhabit. For instance, on some of New Zealand's offshore islands, soils have become less fertile as a consequence of rats destroying seabird colonies since seabird dung has fertilising effects. Pests also negatively impact New Zealand's economy. Mustelids and possums threaten cattle by carrying bovine tuberculosis; rats eat a range of crops, from cereals to fruits; and rogue possums consume more than 7.6 million tons of vegetation each year (European settlers brought the Australian marsupial to New Zealand in 1837 to start a fur industry and, during their 1980s apex, 50–70 million possums roamed New Zealand). Tourism is also impacted by the presence of invasive species, with international tourists seeking out predator-free areas (by way of example, visits to Tiritiri Matangi island increased threefold in the decades after the island cleared its shores of non-native animals). Although it's impossible to put a precise figure on the economic damage caused by non-native species, several experts estimate that pests cost New Zealand up to NZ$1.83 billion each year.[7]

In 1874, an unassuming man named Richard Henry emigrated to New Zealand at the age of 29. Born in Ireland but raised in Australia, Henry was forced to confront life's harsh realities at a young age. His mother and infant brother died during the crossing from Ireland; his father lost and later recouped the family fortune; and Henry worked numerous odd jobs, from carpenter to shepherd to taxidermist.

Wildlife was his great love.

His formal education was limited, but as a young man Henry observed the natural world with a scientist's eye. He spent time with Native Australians, learning about climbing and hunting, and the outdoorsman learned to sail. Why he moved to New Zealand remains a mystery, but historian John Kenneth Hill suspects it had something to do with the turbulent years that preceded his departure (another brother died in an accident, and Henry married only to divorce within two years). He was seeking – and soul-searching – that much is certain.

I felt an immediate kinship with Henry when I came upon his name while reading about the history of conservation in New Zealand. Both of us spent our 20s stumbling as we chased personal and professional stability, ultimately choosing to build lives abroad. Neither Henry nor I became scientists, but we both experienced a powerful yearning to protect wildlife after observing nature with childlike awe. For each of us, writing served as an entry point to conservation: I left law to pursue conservation journalism at the age of 30, and Henry began writing natural history articles for a now-defunct weekly newspaper, *Otago Witness*, in his mid-30s. For the Irish-born Kiwi, depression was a frequent, sometimes intolerable burden; the dark cloud also follows me.

The comparisons end there. Henry became what I call a 'real conservationist' while I work as a translator of sorts, a storyteller who shines a spotlight on people like him. Like all of the human characters in this book, he rolled up his sleeves and put his passion to work even though it meant risking life and limb. Henry pioneered approaches now used by veteran biologists, but he died thinking himself a failure – and in a way he was, but not by any fault of his own. Henry was simply ahead of his time.

Towards the end of May 1895, Henry climbed into his sailboat and began a perilous journey with precious cargo. On his boat were cages containing critically endangered birds, most of them kākāpō, the world's only flightless parrot (its name, which combines the Māori words for parrot and night, reflects its nocturnal nature). The self-made conservationist and his dog had trapped the birds themselves, and Henry had learned how to care for them during months on the mainland. It was finally time to move the endangered birds to their new home, Resolution Island – an offshore, predator-free haven – where Henry hoped they would breed and ultimately proliferate.

The mission was urgent.

Like dozens of flightless bird species, kākāpō started to decline after Polynesians colonised New Zealand with rats and dogs in tow. People targeted the parrot for its feathers and meat, dogs preyed on birds, and Polynesian rats, though small, feasted on kākāpō eggs and chicks – a relatively easy feat given that kākāpō nest on the ground. Like moas, New Zealand's flightless parrots are unable to fly away from danger; besides, they freeze when struck with fear, a strategy they use to evade their only natural predators, birds of prey, which hunt from the skies. But unlike moas, kākāpō survived Polynesian settlement, and even remained abundant in areas that were not intensively settled.

Nevertheless, kākāpō had already vanished from large chunks of their historic range by the time Europeans started moving to New Zealand during the first half of the nineteenth century, and things were about to get much worse for the parrot (and numerous native species). Polynesians unleashed dogs and small rats on the island nation while utilising, and sometimes exhausting, its resources, but they never sought to fundamentally alter the land to which they laid claim. Europeans, on the other hand, colonised new places with the aim of creating 'Neo-Europes – landscapes altered to remind settlers of their lands of origin and to enable them to generate livelihoods from the kinds of extractive or agricultural activities with which they were familiar'.[8] In New Zealand's case, European settlers, most of whom were from the United Kingdom, brought animals, plants and practices designed to build a 'Britain of the South'.

Which explains why many of us, when conjuring visuals of New Zealand, see verdant pastures dotted with sheep instead of ferns and moas. Look closely, and you'll notice pheasants slinking through the country's manufactured farmlands and blackbirds darting between hedgerows. Rabbits are there, too, as are small but mighty foreign hunters – house cats, rats, stoats and weasels. At first glance, the story of how these invaders ended up halfway across the world from their native soils is complex. They were introduced for different reasons and at different moments in time. Rabbits were first brought to New Zealand by whalers in need of meat and fur, whereas pheasants came over for sport. Cows and sheep arrived as Europeans converted forests and tussock to English grasses, which now cover 40 per cent of the country. Norway rats were introduced accidentally after the rodent stowed away on the vessel James Cook used when mapping New Zealand in 1769 – an endeavour that paved the way for European settlement.

Many plants and animals were introduced to remind settlers of home during a time when 'acclimatisation societies' were prevalent. Knowing the consequences, as we do now, it's easy to criticise the fad, but if you had moved halfway across the world to a foreign land in a time before digital communication, you probably would have brought a slice of home with you. Widespread British settlement of New Zealand began in 1840, and in the 50 years that followed, more than 180 species of exotic animals and plants made their way to the Britain of the South. 'So drastic was the impact of European settlement,' wrote historian Michael King, 'that one geographer was moved to note that human-sponsored modification of landscapes which had taken place over twenty centuries in Europe and four in North America had occurred in New Zealand in only one century.'[9]

Although I make no attempt to quantify the damage caused by different classes of invasive species, one invader stands apart in scope and absurdity. When it comes to the introduction of mustelids – a family of small carnivores that in the Kiwi context includes ferrets, stoats and weasels – it's impossible to look at past mistakes forgivingly. For one thing, mustelid introductions were deliberate and thus preventable. The risks were clear, the proposal widely debated. Farmers wanted to unleash the carnivores to control rabbits, which were causing damage to sheep pastures, but scientists like Charles Darwin and Alfred Newton knew that doing so would decimate native wildlife.

Newton called the proposed solution to the rabbit boom 'disastrous', writing, 'No person ... can for a moment doubt that what remains of this [native] fauna will absolutely and almost instantaneously disappear.'[10] Darwin agreed. He visited New Zealand in 1835 and later said, 'If all the animals and plants of Great Britain were set free in New Zealand, in the course of time a multitude of British forms would become thoroughly naturalised there, and would exterminate many of the natives.'[11] Critics made economic arguments as well, explaining that grain and cereal farmers would suffer losses when weasels killed the birds that keep crop-eating insects in check. In the end, the pro-mustelid camp won, and by 1886 more than 8,000 ferrets, weasels and stoats were running free.

It didn't take long for citizens and lawmakers to realise that mustelids weren't just 'natural enemies' of rabbits – the argument that justified their arrival – they also preyed on birds, both native and exotic. The latter especially troubled *Pakeha*, the Māori term for Kiwis of European descent, who couldn't bear to part with pheasant hunting. Thankfully, the protection of native birds slowly became a priority as New Zealand-born Europeans built connections with their homeland. A growing recognition that the country's wildlife was unique – and severely imperilled – further pushed the needle towards conservation.

By the early 1890s, conservationists like Henry realised that kākāpō, kiwi, weka and other flightless birds didn't stand a chance against the onslaught of introduced predators roaming the North and South Islands. The most vulnerable would have to be moved to predator-free island sanctuaries where they could rebound – a tactic so common these days, it's easy to overlook its inventive beginnings. One of the largest offshore islands, Resolution, was selected and became the first island sanctuary in the world (two other offshore islands, Little Barrier and Kapiti, became island sanctuaries soon after). Named for the vessel Cook captained during his second voyage to the Pacific, Resolution is the largest island in a southwesterly portion of the South Island called Fiordland and the seventh biggest in New Zealand. Most importantly, Resolution wasn't occupied by people, and its shores were considered predator-free. Whether it would remain that way wasn't solely down to human choice, however. Hunters frequenting the island could introduce foreign animals at any time, but another possibility – that invasive species could get to Resolution all by themselves – stoked fear among conservationists.

Two years before Henry ferried kākāpō between the South Island and Resolution, he fell into a deep depression. The year was 1893 and he was 48. The self-taught naturalist felt underappreciated, and his career appeared to be dead in the water. Henry had been sounding the alarm bell on declining flightless bird populations since the early 1880s and knew it was only a matter of time before mustelids caused species like kākāpō to go extinct. Then, in 1891, things started to look up when the government declared Resolution Island a wildlife reserve. Henry hoped to become the island's caretaker and bought a sailboat, perhaps prematurely, in anticipation of his appointment, but months dragged on, and no offer came. During this period, Henry's theories about kākāpō breeding, all of which stemmed from his observations in the field, began to crystallise, but the scientific community wasn't interested in his ideas (70-odd years would pass before scientists confirmed that kākāpō don't breed every year and males use a 'track and bowl' system to attract mates, both assertions put forth by Henry). These perceived rejections accrued, causing Henry to feel utterly hopeless. He left his home in Fiordland, travelled to Auckland and tried to take his own life. Miraculously, he survived a self-inflicted gunshot to the head and his spirits lifted soon after when he learned that he had secured the Resolution Island job after all.

He was appointed in 1894 – three years after Resolution received protection – and started translocating and releasing ground-dwelling birds a year later. The process was complex, and without a blueprint to guide him, Henry had to feel his way through every step – from capture to release and everything in between. Working with a muzzled dog, he learned how to locate birds – a notable achievement given the kākāpō's solitary nature and large range. He also figured out how to keep birds alive during periods of captivity that sometimes ran long when inclement weather forced him to delay journeys from the South Island to Resolution. He kept meticulous records, many of which contained groundbreaking information about kākāpō. According to biographer John Kenneth Hill, Henry 'bombarded' scientists and prominent citizens with his field observations and scientific theories.[12]

The middle-aged steward was, in pretty much every regard, prolific, motivated by what Hill calls a deep love of New Zealand's native birds (kākāpō were his favourite).

Henry doggedly searched for another of New Zealand's flightless birds, the takahē, and wrote a book (*The Habits of Flightless Birds of New Zealand*) during the same period that he relocated hundreds of kākāpō and kiwis from the mainland. Doing so required 'sailing his dinghy on waterways and fiords characterised by dangerously violent and unpredictable winds' and manoeuvring through 'nearly impenetrable rainforest'.[13] Over a period of four years, Henry moved 572 kākāpō and kiwis from the mainland to Resolution, at which point he deemed the island full. He then pivoted, spending his time monitoring released birds, writing, and educating visiting hunters and tourists about the island's protected status.

But no degree of protection made Resolution impenetrable. In 1990, mustelids started turning up on the island's shores, confirming what naysayers feared from the start: the distance between Resolution and the South Island was too small to keep crafty mammals from swimming to the newly formed sanctuary. Henry attempted to trap the invading predators, once spending 91 days in the bush looking for a single stoat. He also moved a small number of birds to nearby islands in the fiord of Dusky Sound, but the challenges facing Resolution seemed to grow with each passing day. Hunters increasingly visited with dogs that preyed on protected wildlife, and birds brought to New Zealand during the acclimatisation craze flew over from the mainland. Blackbirds, sparrows and other common English birds ate the food kākāpō and kiwis needed to survive. Henry watched in horror as kākāpō became emaciated and other native birds disappeared, writing in 1907, 'The new berry-eaters can fly and gather up to take the best of everything.'[14]

Henry left Resolution in 1908, by which time he'd moved more than 700 birds to island sanctuaries in Dusky

Sound, and became the ranger of another offshore reserve
on the island of Kapiti. He guarded Kapiti for four years
but was, in his biographer's words, 'too old to cope'[15] with
the barrage of challenges confronting New Zealand's
ground-dwelling birds. At the age of 67, he moved back to
the mainland, his pioneering work in the rear-view mirror.
In 1929, Henry passed away at the ripe age of 84. He died
an isolated and uncelebrated man, his obscurity evidenced
by the fact that only one person – the local postman –
attended his funeral. For a time, Henry was forgotten, but
30 years later, in Act II of the kākāpō conservation saga,
the tenacious naturalist finally received the recognition
he deserved.

For half a century, little was done on the kākāpō
conservation front. The parrots Henry released on
Resolution Island perished, the North Island population
went extinct and those clinging to life on the South Island
retreated to remote, hard-to-reach places like Fiordland
in the south-west. According to the late conservationist
Don Merton,* 'the case was all but lost' when the New
Zealand Wildlife Service made kākāpō a conservation
priority in the late 1950s.[16] However, there was reason for
optimism: locals occasionally reported seeing kākāpō;
and the elusive takahē, long presumed extinct, was

---

* Merton, who died in 2011, was legendary. He is best known for
saving the black robin from extinction using translocations and a
breeding technique called 'cross-fostering': when black robin
numbers fell to five, Merton and his colleagues began moving black
robin eggs to nests built by a more secure species, the Chatham
tomtit. The tits reared the alien brood, unaware that they were feeding
chicks belonging to another species. Meanwhile, the one surviving
black robin pair started anew, successfully laying fresh eggs that
developed into healthy hatchlings. New Zealand is now home to an
estimated 230 black robins and the population is said to be increasing.

rediscovered in 1949. The *Otago Daily Times* said of the exciting news, 'If the Lake Te Anau country could conceal for 50 years a bird as big as a takahē, enthusiasts feel that it may have moas, too.'[17] Moas were long gone, of course, but kākāpō search missions launched by the New Zealand Wildlife Service in the late 1950s eventually started turning up birds.

There was a problem, however, and a mystery. All of the 220-odd kākāpō discovered during the 1960s and 70s were male. The species had no future, not until 1980, when Merton and his springer spaniel came upon five females on Stewart Island that were part of what he called a 'relatively strong breeding population'.[18] Jubilation ensued, but the 'reprieve for the species, and for the New Zealand Wildlife Service, was short-lived', thanks to feral cats. In a span of two years, 70 per cent of Stewart Island's kākāpō were killed by cats, leaving Merton no other choice but translocation. Like Henry before him, Merton began the process of moving survivors to predator-free islands: Codfish, Little Barrier and Maud.

While critical to the kākāpō's recovery, the three islands only offered a combined area of 4,537 hectares (11,211 acres), making the need for additional predator-free sanctuaries more urgent than ever. By this time, conservationists had a firm grasp on the kākāpō's quirks. Merton was certain, for instance, that kākāpō don't mate every year and only breed following heavy fruiting seasons. Merton laid to rest long-simmering doubts about the veracity of Henry's theories and acknowledged the profound impact Resolution's guardian had on conservation, praising, among other things, his 'extraordinarily accurate powers of observation'.[19]

Knowledge about the kākāpō's habits helped Merton develop new conservation strategies, but in the absence of pest-free sanctuaries, such strategies were doomed to fail. Kākāpō wouldn't breed in captivity, making *ex situ* conservation an impossibility. Fortunately, by the 1980s, New Zealand had spent decades utilising ground-based approaches to remove pests from islands. The first significant offshore eradication occurred in 1988 on 170-hectare (420-acre) Breaksea Island. Over the course of 21 days, rats fell victim to a poison called brodifacoum that was placed in 743 bait stations. Traps are another common tool. You'll recall that Richard Henry set out traps to catch the mustelids that invaded Resolution during his tenure, and traps are still frequently used today – for mustelids, especially – but the game changed in the 1990s when the Department of Conservation (DOC) started dropping poison from the sky.

Inspired by agricultural fertiliser technology, the DOC began using helicopters to aerially deliver a synthetic form of sodium fluoroacetate called '1080' across entire islands. By doing so, it substantially scaled up eradication efforts and began to believe – for the first time – that large tracts of land could return to their natural state, at least on offshore islands. By 2013, 14 species of mammal had been removed from more than 100 islands (whose combined land totalled 45,000 hectares (1.1 million acres). As of early 2021, 117 offshore islands had been cleared of predators and New Zealand's Predator Free 2050 movement – which aims to rid the country of all invasive predators by the middle of the century – is actively 'suppressing' predators (keeping numbers very low) on more than 1 million hectares (2.5 million acres).

Although 1080 is a synthetic version of a natural chemical emitted by plants in countries like Australia to repel hungry animals, not all Kiwis approve of using toxins to manage invasive species. Pets have died from ingesting the toxin (dogs are more vulnerable than cats); hunters dislike its impact on introduced deer; and concerns about wider environmental impacts hinder its acceptance. Even though 1080 targets mammals, the toxin can harm birds and has even led to the deaths of endangered species (e.g. kea, another of New Zealand's parrots).

Predator Free 2050 believes that such losses, while problematic, are tolerable in light of the alternative. Animal rights groups like the Royal New Zealand Society for the Prevention of Cruelty to Animals oppose the use of 1080 and other poisons on the basis that they cause 'intense and prolonged suffering'.[20] The search is on for gentler alternatives. Humane synthetic toxins have been trialled and some conservationists are seeking solutions in nature. Environmentalist Tame Malcolm is in the process of interviewing Māori communities to gather information about traditional pest-control practices. On the other side of the spectrum are bioengineering advocates eager to deploy lab-crafted weapons against pests. Daniel Tompkins of Predator Free 2050 says it's a wait-and-see situation. Ethical concerns aside, Daniel says there's no evidence (yet) that gene-editing would be useful in mammals. Brent Beaven – who manages the DOC's Predator Free 2050 programme – isn't pinning his hopes on bioengineering or any yet-to-be-developed technology. 'Having worked in conservation for 30 years,' says Brent, 'I've learned that you have to constantly develop new approaches and tools.'

They may not be silver bullets, but advancements in remote sensing and trap design are allowing more efficient eradication in areas where 1080 isn't an option due to safety concerns. When it comes to large and remote areas, however, all signs point to 1080 being the most cost-effective method. Even so, additional trapping and ongoing monitoring are often required to fight off reinvading predators. According to conservation organisation Forest and Bird, 'well-managed aerial 1080 operations can reduce possum and rat numbers by more than 95 per cent over large areas of rugged and inaccessible country'.[21] Indeed, 1080, which is typically delivered via aerial drops, has been used to clear more than 100 offshore islands and small mainland sanctuaries.[22] Blitzkriegs like these are great at killing pests like stoats, rats and possums – especially on islands, where natural barriers stave off reinvasion. But the mainland and islands within swimming distance are a different story. In such places, the threat of reinvasion is constant – especially from rats.

Here, knowledge is power. Richard Henry was astonished when stoats made their way to Resolution Island. The shock of it thrust him into an emotional tailspin, but even if he had remained level-headed, he wasn't equipped to manage the invasion. Conservationists of today are armed with knowledge and a suite of tools that allow them to combat pests offensively and defensively. Take Resolution – for decades, the island was left to its own devices as deer, mice and stoats ran amok. Then, in 2008, the DOC decided to tackle Resolution's pest problem. Rats, thankfully, never made it to the island and mice are not a current priority, so the team set its scopes on deer and stoats. As you might expect, eradicating the two animals required tailor-made approaches. An extensive network of traps brought stoat levels down, and rifle hunting keeps deer numbers in check. By way of example, 70 stoats were trapped in 2008 and 2009; no more than two were trapped per year between 2010 and 2017.

In 2011, three years after the government's conservation department started eradicating vermin from Resolution, the island became a sanctuary for critically endangered birds for the first time since the 1890s, when 60 endangered mōhua, or yellowheads, were translocated from the South Island. Later, the critically endangered South Island saddleback (whose population dipped to 36 in 1964) returned to the area, and Resolution is preparing to welcome back kākāpō even though its shores haven't been fully cleared of stoats.

Scientists now know that stoats can swim for up to two hours at a time and travel more than 2km (1.2 miles), so reinvasion will undoubtedly occur. To keep stoats at bay, rangers set traps on the mainland and on stepping-stone islands that lie between the South Island and Resolution – a prophylactic strategy. They also closely monitor Resolution, paying special attention to known reinvasion points, and have traps laid across the entire island to catch unwanted arrivals. The biggest challenge? Resolution's large size and unforgiving topography makes finding small, intelligent pests a bit like looking for a needle in a haystack. In 2020, an unusually high number of stoats made their way to Resolution following a heavy 'mast year'. During such periods, trees and shrubs produce more fruits and nuts than usual. The resulting bounty caused rat and stoat populations to rise.

Conservationists have long wondered how sensitive birds like saddlebacks would react to a mast-induced predator boom. In fact, the saddleback reintroduction to Resolution Island was partly inspired by the need to answer this question, says the island's lead ranger, Pete McMurtrie. For a time, Pete and his colleagues were optimistic. In the months that followed the saddleback's 2017–18 reintroduction, birds dispersed and bred despite living alongside a small number of stoats. Then came the 2020 mast and stoat deluge. Since then, Pete hasn't come across a single saddleback.

'While this isn't the news we were hoping for, we now have a fairly definitive answer to our question about whether saddlebacks can live alongside stoats,' he explains. 'But the good news is that the saddlebacks we simultaneously transferred to nearby Pigeon Island are thriving.' Pigeon is predator-free. Resolution isn't a failure, underscores Pete. 'Resolution's large size and limited pests make it a safe haven for numerous species, and conservation gains have occurred in recent years.' Rock wrens have begun turning up, for instance, and the yellowhead reintroduction remains a success.

Offshore, predator-free islands continue to be critical to conserving species like kākāpō. But not all conservation islands are literal. In New Zealand and beyond, fenced mainland reserves are akin to islands in that they provide barriers between protected species and whatever threatens them. We see this, too, in fortress-like national parks and reserves where electric fences separate animals and people.

Perhaps you're wondering how island-like protected areas fit into rewilding given its focus on reconnecting fragmented cores using corridors. If isolated patches (or islands) are problematic, shouldn't rewilders avoid creating them? The reality is that conservationists can't roll out the Three C template in every corner of the world – nor should they. Rewilding is context-dependent. 'Rewilding still has a critical role on islands,' several scientists wrote in 2019, however, on islands the focus is not on 'trophic complexity' but rather on eradicating invasive species.[23] Once an island is free of invasive plants and animals, native species can resume natural behaviours and interactions. In turn, 'original ecosystem functioning' is restored.[24]

Conceptual considerations aside, kākāpō would vanish if every one of the 200-odd parrots were set free on the mainland tomorrow. 'They would not survive long,' says Andrew Digby, a scientist who works for the DOC's kākāpō and takahē conservation programme. 'Cats, dogs, ferrets and stoats would kill the adults, and if any survived long enough to nest, their eggs and chicks would be eaten by rats,' emphasises Andrew. The mainland contains too many invasive predators: too many cats, stoats and rats. It's as simple as that.

But New Zealand isn't prepared to accept its altered reality forever. The country isn't content with the status quo of using offshore islands as ecosanctuaries – an approach some scientists call 'conservation arks'.[25] And why should it be? As evidenced by their Kiwi nickname, New Zealanders strongly identify with their country's natural heritage, but if the 'real' New Zealand only exists on small offshore islands, many of which are difficult if not impossible to visit, the mainland loses much of its biological and cultural value. People, meanwhile, are unable to experience New Zealand as it was, creating a disconnect with profound consequences. Can we really expect people to care about conserving plants and animals they never have the chance to see, smell, hear or touch? An analogy from the art world helps us answer that question: imagine that a world-famous painting (the *Mona Lisa*, say) were taken off display and locked up in storage. Removing the painting from the public's view would alter its cultural, financial and historic worth. Now let's assume the painting has practical value, too – maybe its presence in the Louvre inspires the planet's next da Vinci, a new artist that in turn influences people from across the world. Similar logic applies to New Zealand's flightless birds. Their worth – whether cultural, environmental or practical – is impacted by their visibility. There's another reason New Zealand needs to expand its island network, and it has nothing to do with cultural value or human connection. Many of the country's offshore conservation parks have reached what is known as 'carrying capacity' (the term refers to the number of plants or animals a place can sustain without suffering ecological degradation). If endangered bird populations continue to increase, as is hoped, conservationists will have no choice but to create new sanctuaries for them. New Zealand is in the process of rapidly scaling up eradication efforts, both on offshore islands and the mainland. As of 2015, 10 per cent of the country's offshore islands had been cleared of invasive predators; that percentage has since dramatically increased, says Brent. The manager of the DOC's Predator Free 2050 programme doesn't have a figure, but tells me the majority of offshore islands have been cleared of pests. Several approaches operate on the mainland. There are fully fenced sanctuaries, semi-fenced ones that use the edge of the peninsula as a natural barrier and unfenced areas where pest control is constant.

Given the country's trailblazing work on offshore islands, it's no surprise that Kiwis have also pioneered anti-pest strategies on the mainland, with specialised fencing a key tool in the conservation toolkit. Thanks to its agricultural history, New Zealand already had considerable experience using fences when conservationists started thinking about mainland ecosanctuaries in the 1990s, but fences designed to restrict the movement of livestock are worthless against New Zealand's small mammals. A specialised barrier capable of warding off burrowing, jumping and wriggling animals had to be invented when conservationists created the first large mainland sanctuary, Zealandia, in Wellington in 1994. Twenty-two conservationists, engineers and scientists collaborated in live animal trials that pitted various designs against the jumpers (cats and stoats), the burrowers (mice and rats) and the expert climbers (possums).

The team ultimately settled on 'the simplest, most robust and easiest to install' fence, one containing 'a curved top hat, wire mesh wall and underground skirt'. Zealandia's 8.6-km (5.3-mile) long fence was erected in 1999 and the sanctuary's 225 hectares (556 acres) were fully cleared of 13 invasive species by 2000. Several incursions have occurred since, but extensive monitoring allows Zealandia's staff to quickly respond to the presence of pests, which in turn protects the 40 native bird species that live in (or pass through) the reserve. Of them, 24 are endemic. These birds share their space with a wide range of amphibians, invertebrates and reptiles – including the rare Tuatara. The 'living fossil' is the only survivor of a prehistoric order of reptiles and, like many of the birds on these pages, it went extinct on the mainland before returning via translocation (to Zealandia, in 2005). Unsurprisingly, Zealandia's rewilding extends beyond animals: rare plants are recovering now that possums are out of the way (the invasive herbivores used to annually consume around 400 tonnes of the area's vegetation). It also extends beyond the reserve's innovative fence. As bird populations recover within Zealandia, they start to leave, creating a spillover effect.

Soon after Zealandia's launch, a much larger mainland ecosanctuary opened in the North Island's Waikato district. Sanctuary Mountain Maungatautari (Maungatautari for 'short') had a head start, having been designated a nature reserve in 1912, but pests weren't removed, or pest-proof fencing put up, until 2006. The fence spans 47km (29 miles), making it one of the longest predator-proof fences in the world – and at 3,400 hectares (8,402 acres) Maungatautari is one of the largest pest-free sanctuaries in the world. A surveillance system runs at all times to ensure breaches are addressed immediately (staff are expected to respond to all breaches within 90 minutes). Because of its large size and

impressive security system, the DOC has selected
Maungatautari for the kākāpō's long-awaited return to the
mainland, which is due to occur in late 2022. Deidre Vercoe
– who runs the department's kākāpō and takahē recovery
programme – can't wait to bring kākāpō back to New
Zealand's core. It was Don Merton's dream, she says, and his
dying wish. Deidre and the team assured Merton they would
carry the torch. 'His legacy and vision drive me,' she says, 'I
can still picture the passion in his eyes.'

Returning kākāpō to the mainland, where they once
dominated, isn't just the *right* thing to do. It's critical to
maintaining the longstanding connection New Zealanders
– both Māori and Pakeha – have to the rare parrot. 'The
connection between people and kākāpō is all but lost,'
laments Deidre. 'In recent decades, only a few lucky
individuals have been able to work with or see kākāpō, but
these parrots are the heartbeat of the land, metaphorically
and literally.' The booming calls emitted by males during
courtship will, she hopes, repair a long-lost connection.

Maungatautari, however, isn't quite ready for the
critically endangered parrot. The reserve has to tweak its
fence so that kākāpō are unable to leave the sanctuary.
Because Maungatautari is surrounded by farmland, escape
would be disastrous, says Deidre. Cats, dogs and ferrets
would get to the parrots in no time. As for the fence, it could
be removed and a new one built from scratch, but Deidre
and her colleagues are seeking a more affordable alternative
that would add double electric wire to the existing barrier.
That way, predators are kept out and kākāpō are kept in.

To ensure it works – and is safe[*] – the team has
been testing the design in a small enclosure on Codfish

---

[*] A veterinarian is on-site during all electrical trials.

Island – the kākāpō's main breeding island. Trials have been successful despite the fact that kākāpō are presumably eager to leave the pen in favour of the wider island, suggesting that the fence will be a success at Maungatautari – since it is 2.5 times bigger than Codfish and far larger than the test site.

Kākāpō are also on the verge of returning to Resolution Island in what will be the first translocation since Richard Henry rowed parrots there 120 years ago. They are scheduled to be released on the Five Fingers Peninsula – a narrow peninsula on the island's west coast – this year or next. As we saw above, pest-eradication efforts launched in 2008 are helping to keep deer and stoat levels to a minimum, but stoats remain a constant threat given their exceptional swimming skills. Beech mast seasons will, of course, present the greatest challenge.

As things stand, the DOC is prepared to accept the mustelid's limited presence. 'It's a bit of a scary step,' concedes Deidre, 'but we are pretty confident that the numbers are low enough to allow the kākāpō's reintroduction.' In any event, the reintroduction is necessary, she says, since all of the predator-free islands are full. 'Until we have massive tracts of land that are predator-free, kākāpō are refugees that have to be highly managed.'

Intensive management takes several forms: at the moment, all kākāpō wear a small, backpack-like transmitter that allows conservationists to monitor their movements, and these tiny transmitters also open up personal feeding stations designed to supplement the kākāpō's diet during periods when nature doesn't provide ample food. Many feeding stations contain hidden scales that measure a bird's weight during mealtimes. That way, scientists kill (or in this case, feed) two birds with one stone, thus avoiding the need for catching and manually weighing kākāpō to check their health.

Kākāpō are sometimes moved between islands to enhance the metapopulation's genetic diversity. Juveniles are typically removed from breeding islands, says Andrew Digby, since they are known to interfere with nests, and juvenile males typically join adult males on a 'bachelor island' free of females (Richard Henry, who likened male kākāpō to gay lotharios, would be chuffed). When males are ready to mingle, Andrew and co move them to breeding islands where the magic happens.

Male kākāpō attract mates through a unique set of rituals, including 'booming' whereby males inflate air sacs in their chests, causing the parrots to comically puff up while emitting far-ranging, low-frequency calls. They don't boom from just anywhere. A number of breeding males dig shallow bowls in the ground in the same area so they can put on a show of sorts. It's called 'lek' breeding after the Swedish word for 'play'. Henry was on to these unique rituals long before credentialled scientists validated his theories. Given their intimate connection to nature, it's no surprise that Māori communities were also aware of this bizarre behaviour, passing stories down the generations about kākāpō assembling in an avian playground known as *whawharua*. Word choice aside, I think we can all agree that kākāpō breeding is the stuff of binge-worthy reality TV. Bring the DOC's matchmaking into the mix and you have a smash hit. I can see it now: *Keeping Up with the Kākāpō* or *Love Island: Kiwi Edition*. Either way, I'm in.

Given the kākāpō's fragile state and small population, Andrew and his colleagues closely monitor – and manage – kākāpō nests. Because females incubate eggs and rear chicks all by themselves while breeding males meet up for play dates, conservationists like Andrew support females by playing a role that's part-bodyguard, part-nanny. Among other things, they check the health of eggs nightly during their first week, and if concerns arise they move eggs to artificial incubators. When doing so, the team places dummy eggs in nests so that kākāpō mothers stay in caregiver mode and are, as a result, ready to rear offspring when newly hatched chicks return to the nest. It's a highly managed approach not unlike the one used on giant pandas in captivity – and there's no denying that it's a far cry from what nature intended. However, intensive management is often required when jump-starting rewilding projects. The question then becomes: when, if ever, can managers step away?

Andrew dreams about a wilder world, one where kākāpō flourish without our help. He knows, however, that now is not the time to walk away. 'If people had never intervened to help kākāpō, the species would probably have gone extinct,' he emphasises. 'And if we didn't step in to boost kākāpō breeding, there would probably only be 65 remaining, versus 205, and the species would be on its way to extinction.' While there isn't a magic number, conservationists are aiming for a viable population of at least 500 individuals – the same target, you'll recall, guiding scimitar-horned oryx rewilding in Chad.

The kākāpō recovery team regularly evaluates acceptable degrees of management and is in the midst of crafting new kākāpō recovery guidelines. In some instances, Andrew and Deidre have already begun slackening the reins. 'Kākāpō had to be intensively managed during the 'rescue' phase,' explains Andrew, 'but as the population grows, and we move birds to more remote islands, we have to shift away from that model and pursue more sustainable management.' Rewilding efforts on Chalky Island bear this out. One of the newer breeding islands, Chalky has a fascinating backstory. The DOC knew Chalky contained suitable habitat for kākāpō, but had no idea whether parrots would breed there. An initial trial launched in 2003 seemed like a failure – no booming, no breeding – so the team translocated all of Chalky's females to another site. Years later, Chalky Island's males built bowls and boomed, so females were brought back in 2020. Although breeding has not yet occurred, Chalky's 25 kākāpō are thriving and Andrew expects to see a new generation following the next breeding season. Confident that Chalky is moving in the right direction, the team has decided to take a step back. 'Going forward, we'll rely more on technology for monitoring the island's kākāpō,' explains Andrew.

Reintroducing kākāpō to Resolution will require a similar shift away from intensive management. Because Richard Henry's stomping grounds will never be fully predator-free, the team has to take on risk, and cede control, when returning flightless parrots to the island. According to Andrew, conservationists will initially conduct routine survival monitoring before turning to the more hands-off Chalky Island approach.

I can't help wondering how Henry would feel if he were alive today. He would be astonished, I'm sure, to discover that his efforts – his failures – weren't in vain. The lonely conservationist would have blushed upon reading Don Merton's glowing recognition of his legacy: 'New Zealand's present success and international leadership in the management of threatened species might not have been achieved without Henry's foresight and commitment...'[26] With such high praise bolstering his confidence, Henry might find romantic love, which eluded him in life as much as the secretive takahē. I bet he'd marvel at New Zealand's global leadership on pest control and the successful eradications taking place on Resolution, but he'd simultaneously take comfort in the fact that offshore islands are no longer the *only* safe space for the flightless birds he fought so hard to protect. You see, Henry always considered island sanctuaries temporary solutions, remarking in 1896, 'On the islands the birds may survive for half a century, and by that time people in every corner of the world will realise their interest and value, and then there will be no fear of their becoming extinct'.[27]

His prophesy came true, of course. Today, endemic species like kākāpō, kiwi and takahē are appreciated in New Zealand and beyond. Their unique characteristics make them popular among animal-lovers, but they are more than curiosities. New Zealand's endemic birds are walking, breathing, booming examples of evolution, of nature's logic – and symbiosis. They perfectly fit their native environment, and vice versa, like expertly tailored suits.

To think, we almost lost them and the secrets they hold, and what a loss it would have been: a world of birds gone forever, relegated to history books and museum exhibitions. Instead, we have a fighting chance to save the birds and the land they've inhabited for millions of years. As Deidre says, the future is unpredictable, but many of New Zealand's most vulnerable birds – black robins, kākāpō, saddleback and takahē – seem to be out of the woods following brushes with extinction. But will they ever live freely? Move, eat and breed without oceans and fences standing in their way? Can Predator Free 2050, which many call 'New Zealand's moonshot', actually work?

Brent Beaven thinks so: 'We have a really good shot at achieving it – it's not some pipe dream. There's a logic and reality behind Predator Free 2050 that says, "If we apply ourselves, we can achieve this." But if we don't get there,' he adds, 'the shot was worth taking because we'll be in a much better position than if we hadn't tried at all.'

Deidre is an optimist, too. 'No one foresaw the progress that's been made since Richard Henry's time. Similarly, we're incapable of seeing what the future holds, but I feel hopeful when I look at places like Wellington, where communities are coming together to create large swathes of predator-free land in the middle of a city.' Predator Free Wellington launched in July 2019. Phase one of the 10-year project focused on clearing one part of the city – the Miramar Peninsula – of possums, rats and mustelids. By December of that year, the peninsula was possum-free and nearly devoid of rats, and mustelids hadn't been detected for nearly two months. Now, much of the peninsula is pest-free and locals have achieved many milestones of their own, including ridding an entire suburb of rats. James Willcocks, who runs the programme, is astonished by its rapid progress, 'I thought it would be ten years before we reached some of the ecological outcomes we've already seen,' he says. Numerous forces are at work, but Wellingtonians deserve much of the credit, says James. Ninety-four per cent of them support the project and many residents actively participate in pest-eradication efforts. It helps that they're already seeing the fruits of their labour as once-rare birds spill into the city from Zealandia.

Kakas – another of the country's parrots – are steadily increasing throughout the city following their 2002 return to Zealandia (and the mainland). The flighted bird is beautiful and noisy, so its presence doesn't go unnoticed. 'We now have a young generation of Wellingtonians who know kaka and care about them,' adds James. 'Because they see the birds, they value them. Kakas become part of their identity.'*

For Māori communities, connections like these are essential, because in Māori culture humans are *part* of nature as well as its *guardians* – a concept encapsulated by the term *Kaitiakitanga*. When nature is lost, culture slips away. 'Our language and culture are inspired by the sounds of nature,' explains Melanie Mark-Shadbolt, who wears multiple hats including chief scientific Māori adviser to the government. 'If we don't protect and restore, we lose nature and we lose who we are,' she adds.

A well-known Māori *whakataukī*, or proverb, states: '*Ko ahau te awa, ko te awa ko ahau*' ('I am the river, the river is me'). Another speaks to the importance of conservation: '*Toitū te marae a Tāne-Mahuta, Toitū te marae a Tangaroa, Toitū te tangata*' ('If the land is well and the sea is well, the people will thrive'). New Zealand's environment will never be as healthy, or as pure, as it was before boats landed on its shores. Moas can't come back and centuries of farming have left an indelible mark on New Zealand's

---

* Unfortunately, human interest in the bird's presence has not been entirely productive. Some locals misguidedly feed their new winged neighbours in hopes of bolstering their well-being. In reality, unsuitable foods like crackers and nuts have caused disease and, in some cases, death among Wellington's kakas. According to a 2017 article in *The Guardian* ('Killing kakas with kindness: New Zealand bird lovers threaten future of parrot'), 80 per cent of kaka chicks monitored by scientists in 2016 died because of food handed out by locals. These unintended consequences serve as an important reminder that rewilding is not risk-free (that and community education and outreach is paramount, particularly in the context of urban rewilding).

landscapes, but rewilding is giving the country a chance at environmental health that conservation on its own cannot achieve. In turn, even the most vulnerable birds will have the chance to rebound, reinstating New Zealand's title as the 'Land of Birds'.

CHAPTER FOUR

The Return

Leno Sierra grew up in Mexico and spent her 20s working as a professional dancer, but she now works as a full-time 'pangolin nanny' in South Africa. A unique mammal poorly understood by biologists, pangolins were, for many of us, entirely unknown until scientists identified them as possible Covid-19 vectors. Pangolins are the most heavily trafficked mammal in the world. Although millions have died as a result of the illegal wildlife trade, the lucky few are rescued and rehabilitated, hence the need for nannies. The story of how Leno landed the improbable role is a circuitous one with a familiar beginning: a lifelong nature-lover, she fantasised about travelling to Africa during her childhood in Cancún, a coastal town that, back then, housed more toucans than tourists. Leno finally ventured to Africa in 2011 at the age of 30 to volunteer with an African wild dog project located at a reserve in South Africa's KwaZulu-Natal province. She fell under the country's spell and vowed to move back one day.

'I finally did, eight years later!' she cheerily proclaims during a June 2020 video call. Leno briefly returned to South Africa once – training to become a nature guide – before her big move in 2019. She'd just spent several months volunteering at a rhino orphanage when the pangolin opportunity fell into her lap.

'I was visiting a friend who lives at the andBeyond Phinda Private Game Reserve when I bumped into an ecologist I'd met a few months earlier,' explains Leno. 'We got to chatting, and he mentioned that Phinda needed someone to help with a new pangolin rewilding project. I was like, WHAAAT? Seeing a wild pangolin through binoculars was on my bucket list, and here's a chance to work with one even though I'm not a vet or zoologist? It was mind-blowing.'

For Phinda, the connection was equally fortuitous. The reserve, which is the flagship property of safari company andBeyond and has a long history of conservation, had just begun returning the Temminck's pangolin (aka ground pangolin) to its vast wilderness several weeks earlier as part of a groundbreaking initiative to reintroduce the species to KwaZulu-Natal.

It's led by the African Pangolin Working Group, or APWG, a non-profit that collaborates with the South African government to rescue pangolins from the illegal wildlife trade and, when possible, release them back into the wild. APWG's staff has extensive experience saving pangolins, sending them to the Johannesburg Wildlife Veterinary Hospital for rehabilitation and releasing healthy individuals. However, the partnership with Phinda marked the first time APWG – or anyone – had rewilded pangolins by reintroducing them to areas where they had gone locally extinct. According to Nicci Wright, a wildlife rehabilitation expert who co-founded APWG and the Johannesburg veterinary hospital with veterinarian Karin Lourens, there's nothing else like it.

Nicci has spent decades rehabbing injured and orphaned animals, and has experience with some 350 species – honey badgers and otters are her specialism – but she didn't encounter pangolins until 2007. 'I remember it well,' she tells me over the phone. 'It was the first time I became aware of a poached pangolin offered for sale. Sadly, trade has exponentially increased since then, with a major turning point in 2010 or 11.'

APWG's chairman Ray Jansen also noticed the spike so, in 2011, he and Nicci came together to launch the working group. Since its inception, APWG has managed the release of more than a hundred ground pangolins across several South African regions, but rewilding didn't come up until several years ago. Phinda was an obvious candidate, the perfect partner, says Ray. 'We look for large reserves because these animals can move incredible distances even though they don't run per se.' Phinda boasts more than 28,000 hectares (69,190 acres), and its conservation managers have unrivalled experience.

No one knows when the ground pangolin vanished from Phinda, but the last historic record is from 1974, so experts believe the species has been locally extinct from KwaZulu-Natal for decades. Pangolins are notoriously shy and their nocturnal nature makes them difficult to see. Could they have gone unnoticed all these years? Charli de Vos, one of Phinda's ecological monitors, doesn't think so. 'Someone would have seen at least one during the past few decades if the species had remained in the reserve.'

Predominantly solitary, the ground pangolin lives in woodlands and savannas and requires an abundance of ants and termites for sustenance; they're picky about the species they eat but consume millions (or possibly billions) of insects each year. Pangolins are considered long-lived, with an estimated lifespan of 20–30 years (no one knows for sure) and they reproduce slowly. As a result, when adults are poached, the species suffers more than it would if pangolins bred like rabbits – or even lions.

Less endangered than its Asian counterparts, the ground-dweller is Africa's most widespread pangolin, with a range that begins in central Chad and dips down to several stretches of South Africa's north and east. Records are scant and pangolins are shrouded in mystery, so scientists don't know the precise limits of the species' historic range in South Africa. Ground pangolins are presumed extinct in several regions, including KwaZulu-Natal. Like so many data points, the species' total population size is unknown. The most recent assessment, which was co-authored by Ray and published in 2019, concluded that South Africa contains between 16,329 and 24,102 adult ground pangolins. How many live in other parts of the animal's range is anyone's guess.[1] However, there's no doubt that the species is experiencing an ongoing and precipitous decline. By 2032, it is projected that numbers will have dwindled by 30–40 per cent.

Various threats – from electric fences to local consumption – challenge the secretive species. Pangolins and wildlife trade are always mentioned in the same breath, and there is no doubt that demand in Asia has pushed African pangolins to the brink. Less well known is the fact that communities across Africa have long utilised pangolins and their parts. In some Zulu traditions, for instance, pangolins are the greatest gift one can bestow on tribal chiefs or elders, which is surprising given that pangolins don't survive long in captivity. On the other hand, in Zulu mythology, a pangolin sighting foreshadows drought; to stave it off, one must kill the harbinger of rainless skies. And, as in Asia, some Africans believe pangolins possess healing powers: in Sierra Leone, for instance, scales and body parts are used to treat a range of ailments, from skin rashes to infertility.

International trade only became a significant problem in the past decade. After depleting their native pangolin populations, Asian traders turned to Africa to fill the void. They may be hard to find, but once spotted, pangolins are defenceless against humans because they roll into a ball, making the poacher's job a cakewalk. In 2010, confiscations at African ports began to exponentially increase and body

parts of Africa's four pangolin species began turning up in
Asia. In 2019 alone, 97 tonnes of scales from 150,000–
160,000 pangolins were recorded leaving South Africa, and
Jansen says only 10 per cent of the actual trade is recorded,
which means that as many as 1.6 million pangolins were
killed across Africa that year to feed international demand.
The African Wildlife Federation puts the number much
higher, estimating that poachers kill as many as 2.7 million
African pangolins a year.[2] Losses of this magnitude
undoubtedly threaten the pangolin's future, but no one can
put a number on the percentage lost.

Looking back now, conservationists can see that 2010
was a turning point – and the year was significant for me,
as well. After three unhappy years of practising law, I hung
up my hat and set out to become a full-time conservation
advocate and writer. The idea had taken root a year earlier
when I travelled to Africa for the first time. Like Leno, I
had been desperate to go since childhood. Game drives on
the ranch gave me a taste of the safari experience, and the
walls of my childhood home in San Antonio were adorned
with wildlife photos taken by my mom during a 1970s trip
to East Africa. Her trip sounds downright dreamy, the clear
highlight a day she spent with George Adamson, the
legendary lion rewilding guru. Annoyingly, she remembers
little of their conversation despite possessing photos of him
walking around his property shirtless and barefoot. I like to
think they talked about the type of rewilding Adamson
and wife Joy pursued – the kind detailed in her famous
memoir *Born Free*.

When I landed in Namibia in late 2010, I knew a fair
amount about the animals of sub-Saharan Africa. Books
and documentaries got me started, and I picked up practical
knowledge during my first volunteering stint in 2009, also
in Namibia, when I helped care for dozens of injured and

orphaned animals, from mongooses to lions. In a remarkably short period of time, legal knowledge slipped from my mind like water through a sieve. The empty real estate was soon filled with far more interesting facts, all to do with animals (shockingly, animal behaviour is more interesting than estate tax). Despite this, pangolins weren't on my radar, not least because their shy nature makes sightings near impossible. They are better known in China and Vietnam, where the 'scaly anteater' has long been on the menu, with its meat considered a delicacy. Its scales, which are made of keratin, the same protein in rhino horn and human fingernails, are frequently crafted into medicinal remedies.

Overconsumption decimated Asia's four pangolin species – three of which are considered 'critically endangered' by the IUCN – and, soon, poachers began targeting African pangolins. WildAid, an NGO whose work focuses on reducing demand for wildlife products in Asia, made pangolins a priority in the middle of the decade after seeing the jump from Asia to Africa. 'We worked with a number of stakeholders to help all eight pangolin species obtain protection under Appendix I of CITES (the Convention on International Trade in Endangered Species) in 2016,' says founder and CEO Peter Knight. 'CITES protection was great,' he continues, 'but we needed something else to make it stick.' Ironically, the final push may come from Covid-19.

As scientists sought to pin down patient zero, some speculated that the novel virus jumped from pangolins to people, and overnight our species began to care about the armoured animal, illegal wildlife trade and wet markets. By the time the world woke up in early 2020, Nicci and her team had successfully released dozens of ground pangolins across South Africa and were on their second

year of collaborating with Phinda and another private
reserve, Manyoni, in KwaZulu-Natal.

Nearly every pangolin rescued by APWG comes out of
the wildlife trade. One exception was a male called
Rampfy, who in October 2018 was spotted beside a road
near the town of Hoedspruit, which sits beside the Kruger
National Park and is 450km (280 miles) north-east of
Johannesburg. Named for the veterinarian who provided
his initial care, the young male weighed a mere 1.7kg
(3.7lb) when rescued. Unlike the first two pangolins
released at Phinda, both of which were set free on wild
terrain a month after they were taken from traders, Rampfy
required months of care. Which is where Leno (and her
unbridled enthusiasm) came in.

You'd be forgiven for thinking 'pangolin nanny' is the
job of your dreams. My younger self would have jumped at
the opportunity. In middle school, I nursed two orphaned
grey squirrel pups back to health before tearfully depositing
them at a wildlife rehab centre after acknowledging that I
was ill-prepared for the responsibility of caring for two
helpless creatures. And yet, after reading *Born Free*, Joy
Adamson's account of raising and ultimately releasing a
lioness into the wild, I fantasised about converting my
family's ranch into a wildlife sanctuary. The ranch was sold
soon after, nipping my dream in the bud, but just think, I
could have rescued animals from roadside zoos while
becoming Joe Exotic's mortal enemy! In Namibia, I did get
to play a small part in rewilding a number of animals – an
experience that, while exciting, laid bare the many
challenges (and risks) that come with wildlife rehabilitation.

There was Sir Roger, a feisty olive baboon who urinated
on my head when I bypassed him to peek into a cheetah
enclosure, and two teenage vervet monkeys that nearly
bit my face off during a feeding. On a blisteringly hot

afternoon, I jumped at the opportunity to take several meerkats on a bush walk, but quickly regretted my decision when I realised how much running was involved.

As it turns out, pangolins also move much faster than you'd think.

'They're like ghosts,' says Ray. 'If you take your eyes off one for a split second, even inside the Joburg clinic, it's gone.' Poachers are only able to find them by bribing villagers who spend enough time in the bush to know where a particular individual burrows. 'Young 'herdboys' who graze their family's livestock on communal lands are easy targets,' explains Ray.

When Leno spoke to Phinda's conservation manager, Simon Naylor, about the position, the former dancer worried that her lack of experience would get in the way. 'No one has this kind of experience,' Simon replied. 'We just need someone keen, passionate and responsible.' Leno was hired on the spot and began nannying the following week.

At 8 a.m. on 23 July 2019, she moved her things into a modest stone cottage at Phinda. Rampfy was fast asleep in a crate in Simon's guestroom, so Leno spent the next few hours sitting outside enjoying the sunshine as she waited for the young pangolin, then an estimated 10 months old, to wake from his slumber. When she held him in her arms for the first time, she felt odd. 'There's no other way to describe it,' Leno says. 'I'd never seen a pangolin before and there I was holding the most trafficked mammal in the world. I kept giggling nervously. I didn't know if I was holding him properly and worried I might hurt him or vice versa.'

For two days, Leno received training from a woman who looked after Rampfy before Phinda assigned him a full-time babysitter. Then, on day three, Leno was set free.

She constantly corresponded with Nicci, peppering her with questions night and day. 'I asked her if he was eating enough, if his bowel movements were normal. When it comes to pangolins, you can't Google this stuff. So few people know anything about them.'

Leno quickly discovered that pangolins don't bond with their keepers. They don't make eye contact and have no interest in cuddling or keeping close. As she puts it, 'There's no interaction at all.' Though Nicci and Karin say youngsters occasionally seek human touch, the rehab expert and veterinarian agree that pangolins are more aloof than other species.

'They're unlike any other animal,' says Nicci. They eat like anteaters and have protective exteriors like armadillos or turtles, but pangolins' closest relatives are felids, the scientific family containing cats. With limited information about the ground pangolin's life history, Nicci and Karin had to improvise when they began treating their first patients. 'We had to learn nearly everything through trial and error,' explains Karin. Today, the hospital is the only South African facility with adequate experience to handle pangolins.

Lack of knowledge aside, the vet had a critical advantage – 14 years of experience treating a wide range of African mammals. However, she lacked guidance when it came to medicinal dosages. Knowing the link between pangolins and felines, Karin initially used cat-tested doses when treating pangolins suffering from conditions like dehydration, emaciation and pneumonia, but it soon became clear that feline doses were inappropriate for pangolins. Bloodwork posed another massive challenge. 'We didn't have a baseline because there weren't any existing panels to run our samples against,' explains Karin. To rectify this knowledge gap, the vet spent several years researching the ground pangolin's chemistry and haematology as part of a master's degree in science. Among other things, she learned that pangolins' red blood cells are closer in size to bovine cells than they are to cat or dog cells. 'My findings have been tremendously helpful because we now have a baseline when conducting an animal's bloodwork, which in turn allows us to create tailor-made treatment plans.'

Karin and Nicci also had to work out what to feed rescued pangolins before the animals were ready to start foraging (because they are seized from poachers, pangolins entering the clinic have typically gone weeks without food or water, so they have to work up to eating solid foods again). Like Daphne Sheldrick, the British Kenyan woman who founded Nairobi's well-known elephant orphanage and developed the milk formula still reviving calves today, Karin and Nicci had to create a pangolin meal plan. They looked at the captive diets of other ant-eating animals, like echidnas, for inspiration, eventually settling on a formula and general rehabilitation protocols. Rescues of all ages are put on IV drips then gradually tube-fed a liquid concoction full of protein.

On day four, staff members take the pangolins out for foraging sessions. These sessions are critical: in captivity, pangolins won't eat of their own volition, so once they move beyond the tube-feeding phase, they need four to six hours outdoors a day to forage on ants and termites. To keep track of them during night walks, 'We attach a flashing light to the pangolin like you'd put on a dog's collar – and off it goes like a flashing disco ball,' Nicci says, laughing.

For a little over three months, Leno spent her waking (and sleeping) hours walking with Rampfy in the bush. Young pangolins are more active during the day than adults, and Rampfy got moving at noon during Leno's first few months, which coincided with winter in South Africa. After breakfast, Leno would sit and wait outside Simon's house. As soon as she heard movement in the youngster's box, she went into the guestroom to retrieve him.

She'd hop in the car, Rampfy tucked into a cat crate in the back, and pick up a Phinda employee named Alfred Ghina who didn't speak a lick of English. Going into the bush alone is never safe, especially after dark, so on night walks Leno's companion carried a firearm. At the start, colleagues pointed Leno towards areas where Phinda's newly released pangolins liked to forage, but Leno was, for the most part, her own boss.

'From the beginning, my job was very independent, and I liked testing out new areas. Because it was still cold out in those early days, the ant and termite nests were hardened. We'd have to break them up for Rampfy – he couldn't get to them on his own – so Alfred became an expert at cracking them open with a pick.'

Pangolins are capable of walking further, and faster, than field biologists ever imagined (a scientist managing releases in Limpopo recorded one individual moving 42km (26 miles) in the span of two days), but Rampfy's age, size and vulnerable condition meant that Leno often carried him between insect hotspots, his body awkwardly draped over her shoulder. For the best chance at a successful release, Rampfy needed to gain weight. 'I was like an Italian mom fattening up her son,' she jokes.

When winter turned to spring, Rampfy's sleep schedule dramatically changed, and before long, Leno found herself having to cajole him out of bed at 6.30 p.m. to ensure they'd have a long enough stroll before the anti-poaching unit punched the clock at 9.30 p.m.

Like Jane Goodall – who discovered chimpanzees' tool-making skills without any relevant credentials (her Cambridge studies commenced *after* she observed chimps using sticks to collect termites from Tanzanian mounds)

– Leno uncovered new aspects of pangolin behaviour. She was bowled over the first time Rampfy rolled in zebra dung – he apparently wanted it coating every scale – and he once made a specific sniffing sound that none of the experts could decode. Leno hypothesised that it was inspired by one of two things – pain or fear – and is now certain fear drives it, having witnessed another pangolin sniffle under similar circumstances. Of course, the finding is merely anecdotal, but as scientists learn more about pangolins, Leno's theory may become scientific certitude.

They may not bond with human caregivers, but according to the experts, pangolins have individual personality traits and preferences. 'Some are so shy, they roll into a ball at the sound of someone stepping on a branch,' explains Karin. 'Others are a bit aggressive and will hit you with their tail.' Pre-rescue experiences undoubtedly influence these behaviours. Cigarette smoke seems to bother all of the pangolins treated in Johannesburg, and most of them prefer women. Nicci and Karin suspect these preferences stem from the fact that many poachers and traders are men that smoke. Rampfy hates the sound of aeroplanes, so Leno never walked him near Phinda's airstrip after noticing the trigger. A youngster Karin cared for liked when she tickled his belly, and when Rampfy became worn out during his initial bush walks in Johannesburg, he'd stand by her leg with a pleading look that screamed, 'Pick me up!'

Typically, conservationists affix tracking collars, tags or chips to animals released into the wild to monitor their status and evaluate the project's success. Phinda's staff is no exception. Indeed, they're highly experienced when it comes to tracking and translocation, having reintroduced the first batch of large mammals (cheetahs and rhinos) in the early 1990s. Since then, 19 additional species – from elephants and hippos to crocodiles and rock pythons – have been reintroduced to the former farm. In Zulu, Phinda means 'the return', a moniker that reflects the reserve's wildlife reintroductions *and* ecological restoration.

Restoration took several forms, but typically involved removing artificial or alien things. Fences were brought down as staff pulled out the cattle guards and livestock troughs used during the property's farming days. Dams, likewise, came out. Phinda boasts seven distinct habitat types, including a rare sand forest, and some needed a boost before native animals were able to return. Bush clearing opened up savannas, and alien plants were actively removed to let native species proliferate. 'Controlled burns have also played an important role in restoring the grasslands,' emphasises Simon, who adds that fire's role in restoration and rewilding is often overlooked – a point echoed by many of the rewilding experts I interviewed.

I visited Phinda in late 2017, nearly 20 years after its restoration began. Before travelling to KwaZulu-Natal, I spent several days at Londolozi, another well-known South African reserve with a fascinating backstory (and links to Phinda). The family that created and runs Londolozi helped establish Phinda and the conservation model that guides it. Over coffee one morning, I chatted with Dave Varty, who ran the company that preceded andBeyond, and I met his son Boyd that afternoon. Boyd's eloquent memoir, *Cathedral of the Wild*, introduced me to the history of private conservation in South Africa and inspired me to see the country's varied landscapes.

After three nights at Londolozi, where I saw multiple leopards and witnessed a pair violently mating, I made my way to Phinda. Not only did I want to soak up the reserve's history while driving through its seven distinct ecosystems, I hoped to meet up with Simon to learn about the team's conservation priorities. Because my trip occurred in November 2017 – two years before the pangolin project got off the ground – pangolins weren't on the agenda when I met Simon on my first morning, but I was ecstatic to learn that two days later Simon and his colleagues were planning to collar an elephant and de-horn several rhinos. Not only that, an Animal Planet film crew was on-site to film the high-stakes procedures, making my stay even more exciting.

After chatting with Simon, I met up with my guide and set out for a leisurely day of game drives. Because I knew a bit about the reserve's history, I looked for signs of its agricultural past, but none appeared. On the other hand, wild landscapes – and animals – were all around me as I explored the sprawling property on game drives. I photographed lions mating for the first time, saw my first cheetah cub and spent a half hour staring straight into the eyes of a confident male leopard closely guarding a nearby kill. Spoiled by bountiful wildlife and diverse landscapes, I completely forgot that, 20 years earlier, the South African paradise would have resembled a Texas cattle ranch.

It was almost as if people had never occupied Phinda: never altered its landscapes nor hunted its large mammals. Although Phinda broadcasts its history, tourists could easily visit the reserve without understanding the role people played in its deterioration *and* regrowth. Had I travelled to the site in a tourist daze, or been less knowledgeable about conservation, I might have come and gone without learning the reason behind Phinda's name. Hearing Phinda's story from guides and marketing materials is one thing, but going behind the scenes, as I did, lets you glimpse the reality of hands-on rewilding.

I tagged along as a journalist (and later published a *National Geographic Traveller* story about the experience), but safari guests are invited to witness conservation in action if their trips happen to align with planned activities like radio collar-swaps. Always innovating, andBeyond figured out that it could raise money (and awareness) by letting guests pay a premium to watch wildlife management in action.

For many conservationists, dehorning and other forms of intensive management are out of the question due to financial constraints, but Phinda, as we'll see later in the chapter, is well-funded. A controversial tactic used to dissuade poachers, dehorning can cost as much as $3,000 per animal. Helicopters are required, as are large teams and specially trained veterinarians. Risks arise any time animals undergo sedation.

Simon didn't just let me observe his team at work, he offered me a prized seat in the helicopter sent on a reconnaissance mission the afternoon before the collaring expedition (Phinda needed to locate the elephant herd to avoid a wild goose chase the following morning). I fancy myself a master spotter, but my eagle eyes failed (and embarrassed) me during the hour-long chopper ride. On several occasions, I tapped the pilot's shoulder to say that I'd seen a lone bull below only to discover that my grand sighting was of a bush, not an elephant.

The next day was the most exhilarating of my life. We set out at 4.00 a.m., a solid four hours before my preferred wake-up time, and snaked our way through Phinda's sand forest, a rare habitat type where we had seen the herd congregating 12 hours earlier. We weren't following a linear path to an appointed location because elephants are often on the move. According to Elephants for Africa, elephants are capable of travelling 195km (121 miles) in a single day, and their renowned capacity for storing long-term memories means they often form negative associations with helicopters because poachers and vets both shoot elephants from choppers (one using bullets, the other tranquilisers). As their blades cut through the sky in a cacophonous roar, trauma can be triggered.

With radio communication and the occasional helicopter sighting guiding us, those of us riding in Land Cruisers swiftly navigated the forest's sandy roads, stopping once to watch lions copulate and again to track down the helicopter team after it flew out of range. By the time we parked near the sedated female elephant, my face was so covered in dirt, I finally had a real-life application for the expression, 'Eat my dust.' If only my vehicle had been first in line.

Watching Simon and his colleagues deftly manoeuvre around the elephant's slumbering body to replace the collar was surreal. For them, the task looked natural, almost ordinary, and I found myself slipping into a dreamlike state, momentarily forgetting that our mission was fraught with danger – for the elephant and us. The lions we'd seen on the road could appear at any moment, and the elephant couldn't safely remain under sedation for long. The second Phinda's veterinarian administered an injection to bring her back to life, Simon sent us running to the car with a wave of his hand while he and the vet remained behind to make sure she safely regained consciousness. I couldn't help wondering: when she reunited with her family members minutes or hours later, would they extend their trunks to hers to communicate affection and comfort? Would they know the leather and metal device hanging from her neck was intended to help, not harm? Research has shown that elephants can remember a range of perceived threats with incredible precision. In East Africa, some herds negatively react to the sounds, smells and colours of Maasai herdsmen since human–elephant conflict often occurs with Maasai communities.

Unlike elephants, pangolins don't require sedation before being fitted with tracking devices, so Simon and his team can easily place VHF and satellite tags on every individual they release. But when the first two rewilding candidates, Louis and Luna, arrived at Phinda in May 2019, tracking devices for the species were non-existent. Simon's team could have pushed back the release date until securing the equipment they needed, but pangolins are highly susceptible to stress and disease while in captivity, so they made the difficult – and potentially life-threatening – decision to track the two without tech's guiding hand.

Doing so meant following the armoured animals around the bush 24/7 for two weeks straight – when the first tags arrived. 'We worked in groups of two doing 12-hour shifts with each pangolin,' says ecological monitor Charli de Vos. Charli and her colleagues rested when the pangolins slept, often twiddling their thumbs outside burrows for hours at a time, and jogged behind their precious cargo when foraging resumed. As time passed, Charli and Simon noticed patterns in the adult pangolins' schedules; they now place the preferred wake-up time between 8 and 10 p.m. during summer months and 11 a.m. and 4 p.m. in the winter.

'Pangolins hate when it's windy,' says Charli, 'and they prefer an outside temperature of 21°C. We've also discovered that their burrows are always this temperature.'

Ground pangolins are equally fussy about their food. Preferences are individualised and may stem from where the animal began its life. Unfortunately, rescuers like Nicci and Ray rarely uncover information about an animal's origin. Did it come from within South Africa or another country, like Mozambique or Zimbabwe? Rescuers can't be sure since trafficked animals are passed around like hot potatoes before landing – either in a market or, when fate intervenes, back in the wild.

'We think a lot of the ground pangolins we intercept come from Mozambique and Zimbabwe,' speculates Ray. 'Poachers capture the pangolin from within their own country, and then come here to sell them since they fetch higher prices in South Africa.' Ray won't share the current price of a poached pangolin, saying only that their value matches that of a rhino horn, since broadcasting such information could encourage poaching.

With no suitable tracking equipment at their disposal, Simon and Charli had to take matters into their own hands. 'At the time, no one had tags that were light enough for pangolins,' says Charli, 'so we began working with a manufacturer to create a pangolin-specific design.' They're now on their fourth iteration, which Charli thinks could be the one. 'We've had to readjust the way the antenna sits on the tag because the antennas constantly break off, either because the pangolin goes into a burrow, causing it to snap off, or because a predator, like a lion or hyena, chews it to pieces.' The pangolin, safely rolled into a ball, is fine. The antenna? Not so lucky.

Because Phinda is owned by luxury safari company andBeyond, an established entity whose major shareholders include members of the wealthy Getty family, the reserve is able to purchase tracking equipment, which can be cost-prohibitive for smaller conservation organisations. As of July 2020, a single satellite tag cost £907 to manufacture and the team had to pay a six-month user fee of £200. What if it stops working and requires refurbishment? That costs £650. At £210 a pop, VHF tags are more affordable and allow conservation managers to locate an animal in real time using a radio transmitter that beeps when in proximity to the tag. But you need to know, even approximately, where to start your search when you're dealing with a huge area like Phinda. That's where satellite tags come into play. They provide hourly location updates – unless the animal's gone into a burrow, in which case trackers have to wait until the animal emerges.

The tech may be imperfect, but it's better than roaming the African wilderness all night long with danger lurking around every corner. Phinda's monitoring team has experienced plenty of heart-thumping moments. On one occasion, an angry hippo charged when a spotlight accidentally landed on him. One guy had a rifle and was prepared to fire, but the woman holding the spotlight lost her cool and ran for the safety of a tree. The others, now blanketed in darkness, followed suit, their feet pounding across uneven terrain until they found a place to hide.

Leno tells a similarly terrifying story: after transitioning to a rehabilitation role at a neighbouring reserve, Manyoni (which is also collaborating with APWG to reintroduce ground pangolins), she and two colleagues went out one night to track a male named Gomo. He was in a particularly thick part of the bush, where thorny branches had to be shoved aside to make headway. 'It got to a point where we felt like we were in a movie,' says Leno. 'One of us said something like, "The visibility is so bad here, we're all gonna die." We'd just picked up Gomo's VHF signal when my colleague Epsy raised his fist in the air, signalling, "Stop!" All three of us froze in place.'

Then Epsy began walking again, indicating that the coast was clear, but moments later he was shouting at the top of his lungs while raising his arms over his head in a wild waving motion. Suddenly, the bushes began to move; an animal was in their midst, but the threesome couldn't see it. Leno threw her telemetry equipment aside and mimicked Epsy, hoping additional sound and motion would frighten off the potential predator. Several feet away, Epsy tried to load his rifle, but Leno could hear the bullet jamming in the chamber. With dense vegetation all around, none of them had clear visibility, even with their torches. Then she saw it, the threat in their midst – a large male warthog, barrelling towards them. Epsy and Leno jumped out of the way, but the warthog grazed the leg of their third colleague. Fortunately, its tusks just missed the man's skin, leaving mud instead of carnage. I know what you're thinking: *The Lion King*'s beloved warthog Pumbaa is harmless, incapable of hurting a fly. In fact, warthogs can inflict serious damage on people. A pair of them nearly killed the woman who runs the conservancy where I volunteered in Namibia after repeatedly thrusting their tusks into her legs.

As Leno points out, members of anti-poaching units are used to walking around the bush at night, but they choose their own routes whereas pangolin-walkers have to follow the forager. And pangolins don't just move faster than anticipated – they're able to navigate rocky, even steep, terrain. Ordinarily bipedal, they use all four legs when rock-climbing and they're very skilled, says Charli. 'One of our females walked 32km, crossed a major highway and ended up in a different reserve. When she got too close to a community, we went to fetch her, but it took us three days to find her because she'd made her way to a very inhospitable area at the top of a mountain.' A Phinda intern who teaches rock-climbing on the side scaled it to pick up the rogue female and later revealed to Charli that if her climbing buddies knew she'd gone up without ropes, they would've killed her.

Since the KwaZulu-Natal project is still in its infancy, protective measures like these are occasionally required, but, like all rewilding projects, this one aims to become self-sustaining. APWG has successfully returned ground pangolins to a number of South African regions (where the species never went extinct) without the need for long-term monitoring. In the short term, scientists track each pangolin's movements using custom-designed VHF transmitters, but some pangolins go out of signal range within a few days, making data-collection a challenge. From October 2017 to November 2019, 41 pangolins were released and tracked, and of those, 71 per cent were alive when their location was last logged. No individuals have been re-poached thanks to the creature's ghostlike nature and APWG's caution choosing release sites. Together, these achievements bode well for the pangolins navigating life at Phinda and Manyoni, but the most sustainable outcome involves the end of the trade. WildAid's slogan, 'When the buying stops, the killing can too,' says it all.

Millions of pangolins have vanished from the wild over the course of a single decade, but without a clear baseline we'll never know the true extent of this loss. However, when it comes to conservation, small gains are everything, and pangolins are poised for a comeback. In the wake of the Covid-19 pandemic, China removed pangolins from the country's list of approved traditional Chinese medicine ingredients and all around the world people started appreciating the animal's perilous state. Meanwhile, in South Africa, pandemic-induced lockdowns brought pangolin poaching to a grinding halt.

How long it will last is anyone's guess, but a hardcore conservation optimist might say the pandemic will forever change the environment for the better. Until then, pangolins like Rampfy and the dedicated monitors risking life and limb to keep them safe are inspiring hope while helping scientists learn more about pangolins, which in turn enhances conservation. Leno, for her part, is content, 'I'm close to bankrupt now, but it's totally worth it. Who else gets to do a job like this?'

CHAPTER FIVE

The Rewilding We Don't See

All those who have chosen to inhabit the San Antonio River valley have had to take into account the flow of water beneath and across the land. Whatever the form of settlement, and by whatever name the inhabitants have called themselves, each group, from the Tonkawa to the Spanish, the Mexicans to Americans, has framed its collective destiny and social structure around its capacity to draw upon available water supplies.

Char Miller, *Deep in the Heart of San Antonio:*
Land and Life in South Texas

As a little girl, my world felt wild. My family lived on a 2-hectare (5-acre) lot in an idyllic part of San Antonio, Texas, called Terrell Hills. Oak trees were everywhere. Birds and insects, too. Back then, I didn't know much about birds, but I wanted to be like them. I frequently stood on the edge of our wrap-around wooden deck imagining myself soaring over the lawn and into the skies. I could fly, I was certain, I just needed to find a way to unlock my avian powers.

When I wasn't attempting flight, I was busy exploring the backyard with my brother Jeff. We had names for the wildest parts of the property, the spots that were (or seemed) untamed. There was 'The Jungle' – a curved stretch of knee-high plants where we pretended to be Spanish conquistadors mapping the New World – and 'The Forest'. The Forest began at the northern edge of our yard and extended beyond the property line. It was undeveloped urban wilderness that belonged, it seemed, to no one.

In our backyard, human-made sounds were few and far between. No buses or trains passed through our neighbourhood, and since our ranch-style house was set back from the street, we rarely heard when cars rolled by. Nature's orchestra, meanwhile, never missed a beat. My favourite performers were the cicadas, a type of insect that, when present, fills the air with a constant buzz. In *The Iliad*, Homer likened several wise orators to 'cicadas that chirrup delicately from the boughs of some high tree in a wood',[1] but I saw them as musicians, not speakers.

Decades later, I still hear cicada songs when I melt into my bed each night. Eyes closed, I picture The Forest, my first wild domain. I travel there in my mind, allowing myself to think like a naïve child once more. Jeff is running ahead of me when he suddenly stops, kneels down and picks up a volcanic rock for inspection. We stash it in the unwieldy fort we constructed from broken tree branches, scrap metal and other discarded things, placing it beside a sparkly geode retrieved weeks earlier. Some of these discoveries still sit in my parents' house. Others were lost like The Forest itself.

Its dewilding happened slowly. The first stretch of shrubs and trees came down as a neighbouring house went up. Later, a second house claimed our fort and the trees encircling it. Finally, a third home stole its last vestiges, snuffing out what remained of our local wilderness.

By the time we moved houses when I was 10, the woods and the treasures it housed were gone. It was a small loss – nothing compared with the global deforestation that's occurred during my lifetime – but a loss nonetheless, one felt deeply by an innocent child who had never witnessed nature's destruction.

What happened across the fence was happening across the city. In 1981, the year of my birth, San Antonio's population stood at 966,000.² Ten years later, 1.2 million people called San Antonio home, and by 2021 the population had risen to 1.45 million.³ Sleepy old San Antonio is now the seventh most populous city in the country (Dallas and Houston also make the list, and Austin clocks in at 11). Of America's 10 biggest cities, San Antonio grew the fastest between 2000 and 2010, and came in second during the preceding decade. On their own, these stats don't tell us much about the city's evolving relationship with nature. If San Antonio were built like Beijing, high-rise apartment buildings could have sprouted up to house the city's growing human population, but, as I'm sure you've heard, everything's bigger in Texas, houses and lawns included.

So, the city ballooned. Like my waistline after several weeks in the Tex-Mex mecca, San Antonio expanded to fill the void otherwise known as nature. I witnessed the changes in real time, stupefied every time a new superstore or interstate came under construction (do we really *need* a fast-food restaurant and gas station at every corner?). Worse, I had a front row seat to San Antonio's suburban sprawl since one of the fastest-growing planned communities arose on land my family sold to real estate developers.

My maternal grandparents (the ones who owned the ranch described in Chapter Two) purchased Greenwood Farms in the 1950s as a weekend retreat. Part of the 445-hectare (1,100-acre) property had been a working dairy farm, and my grandparents raised cattle and show horses on-site, but the land was largely untamed, as was the wider area. Like much of the Texas Hill Country, the region's landscapes are a sea of rolling hills, oak and mesquite trees, and exposed limestone that dates back to the Cretaceous Period – when the region was frequently underwater.⁴ The many fossils I unearthed allowed my young mind to grasp, just barely, the painstakingly slow evolution of life on Earth, but I gave little thought to the future. In school, I learned about the extinction of dinosaurs, but extinction seemed a far-off, abstract concept, the sort of thing that happens to *them*, not *us*.

During the first six months of my life, my family lived at Greenwood Farms, but even after we moved into the house with the wooden deck and imaginary jungle, we frequently retreated to the farm, whether for long weekends, school breaks or holidays. The farmhouse, part of which is 120 years old, was a unique homestead that contained enough nooks and crannies for a lifetime of hide-and-seek, but Jeff and I spent most of our time outside.

We pedalled our bikes across the farm's gravel roads and proudly wore the scars that resulted from underestimating sharp turns. During the summer, our parents walked us down to a series of caves where thousands of female bats roosted while rearing their pups. As night fell, hungry mothers emerged for their nightly feast, their winged bodies casting shadows on the ridge where we watched them zigzag through the air. We named our favourite lookout point 'Picnic Hill'. There, we built campfires, competing to see who could get theirs going first, and who could keep theirs alight the longest, often staying until dusk. Cedar trees, I learned, were great for building campfires, but loathed by farmers and environmentalists. (*Juniperus virginiana* is native to Central Texas, but the species hogs water, fuels wildfires and unnaturally dominates disturbed landscapes, becoming, as one writer put it, 'the vilest plant thriving in Texas'.[5])

I was born in May 1981 and lived at Greenwood Farms from the end of that month until late autumn of that year. That summer, for reasons I still don't fully understand, my grandparents sold 95 per cent of the property to real estate developers. They held on to the farmhouse and 28 surrounding hectares (70 acres), but the rest of the land became a posh gated community called The Dominion.

Unlike Greenwood Farms, The Dominion was an appropriate moniker for the property given how quickly (and fully) the developers exerted control over it. Within several years, a golf course sprung up on a stretch of land that abutted our fence line. We carried on visiting the bats, but now we stood on a manicured green instead of a rocky ridge. The community's first residence was completed in 1984 and a string of others followed, as did a central clubhouse where members could eat, swim and play tennis. We were given complimentary membership as part of the sale, and for a time, I lapped up the club's amenities without considering their cost.

Greenwood Farms and The Dominion may have been strange bedfellows, but for a decade the two properties coexisted. It wasn't until I was in high school that we began to feel The Dominion's pinch. Until then, houses had only gone up in areas we couldn't *see* from our property. Now they were encroaching on us, their Spanish-style roofs visible from Picnic Hill. Soon, houses were so close, we could hear people coming and going from their new homes, so we stopped going to the hilltop. It wasn't gone, not like The Forest, but it may as well have been. In any case, it's gone now. My mom's generation sold the remainder of the farm after my grandmother's passing in April 2000.

The Dominion opened my eyes to urban development and it undeniably encouraged the rapid development of San Antonio's northern fringes. It anchored them, too, by bringing people (and money) to what might have remained unknown farmland. As I said at the start of the chapter, I had a front row seat to San Antonio's growth and related dewilding, but I didn't appreciate the scope and rapidity of the metamorphosis until Google Earth released its Timelapse feature.[6] A quick scroll through satellite images taken between 1984 and today presents a depressing bird's eye view of the changing metropolis. Much of the green on the map – The Forests and Picnic Hills – turned white as houses, parking lots and strip malls consumed the city's lovely landscapes. This growth simultaneously strained the city's natural resources, of which water is paramount. Water drew indigenous people to the region thousands of years ago and prompted Spanish colonisers to create what became one of the crown's most important outposts. Ever since, residents 'have had to take into account the flow of water beneath and across the land'.[7]

Several years ago, a simple thought flashed through my mind as I downed a glass of cool tap water on a hot summer day. Water, I thought to myself, is nourishing, delightful and free. *Free*. As soon as the word hit my brain, I realised my mistake. Water isn't free – not in the monetary sense nor the wider one. Nor is water wholly delightful. Sure, we need it to drink, grow crops and bathe, but water also wreaks havoc, as San Antonio knows all too well. In the semi-arid zone prone to flash-flooding, water is both friend and foe. Water, moreover, is central to the city's identity.

San Antonio is inextricably linked to an eponymous river snaking through its core. Fed by several springs located 6km (4 miles) north of downtown, the San Antonio River has been used by people for at least 11,000 years. With an average width of 14m (45ft), the 386-km (240-mile) long waterway is relatively small, but beauty and plenitude make up for its modest size. When Spaniards turned up in the late seventeenth century, they discovered a group of

Payaya Indians living in an area they called 'Yanaguana', which in their language meant 'refreshing waters'. As we see in Chapter Seven, Argentina's rewilding features a place given a similar name, one signifying 'shining waters'.

Spanish explorers crossing into present-day Texas from Mexico 'measured a day's march by the distance between water holes, springs, creeks and rivers', and the San Antonio River's location in the northern reaches of New Spain made it a critical refuelling station between Mexico and end points in the American West and East Texas. The city wouldn't remain a way station, however. Ample resources and a strategic location lent the area to permanent settlement. A Spanish friar was smitten, calling the would-be city 'the best site in the world, with good and abundant irrigation water, rich lands for pasture, plentiful building stone, and excellent timber'.[8] A critical water source, the river was paramount. As historian Mary Ann Noonan Guerra noted, 'Because of the river the city was born.'

Spain didn't put down roots until 1718, but the city was given a name as soon as it was 'discovered', with San Antonio selected in honour of the 1691 feast day of San Antonio de Padua, a celebration that coincided with the first documented arrival of Europeans. Beginning in 1718, envoys of the Spanish Crown began the process of altering San Antonio's natural features to accommodate the locale's first agricultural society. Irrigation ditches, or *acequias*, were dug parallel to the river as colonists harnessed its resources ('With water came power,' explains environmental historian Char Miller, and 'with water came growth'[9]). Simultaneously, work began on the Spanish missions for which San Antonio is famous. The five built along the San Antonio River were strategically located: a midpoint between the nearest Spanish settlement (in what is now the Mexican state of Coahuila) and

Spanish missions established in East Texas. Mission complexes, or presidio-missions, served multiple functions – from religious conversion and worship to agricultural training grounds – and shaped the growing city, both literally and culturally.

The Alamo, which was originally called San Antonio de Valero, came first – in 1718 – but had to be relocated a year later due to flooding. Flooding, it soon became clear, posed a recurring, seemingly unavoidable threat to the public's safety. Needless to say, the San Antonio River began breaching its banks long before Europeans turned up, and people have utilised the waterway for thousands of years. How did indigenous communities manage? The answer boils down to lifestyle. Because such communities moved with the seasons and weather, and didn't have crops or permanent homes tethering them to particular locations, they were able to use San Antonio's natural resources when doing so made sense. When it didn't, they went somewhere else. Spaniards interacted with (and converted) many members of these indigenous communities, presumably learning from them. In light of this exchange, you might be wondering why San Antonio's founders decided to build the city on a floodplain.

In part, they had no choice. To survive, they had to grow crops, and to grow crops they had to live and farm near water. 'The Spanish couldn't manage the city's flooding issues because of how they produced food,' underscores Char. They took a calculated risk. 'The city's Spanish planners were probably thinking, "It floods but not every year, and we can rebuild after major catastrophes."' Wise or not, that logic endured for nearly 200 years. In some parts of the United States – Houston and New Orleans come to mind – leaders continue to put off for tomorrow what they can do today.

But there's another factor at play, and it relates to a series of regulations that governed the Spanish conquest of the New World. Guided by the Laws of the Indies, Spanish emissaries were asked to populate sites that contained 'good and plentiful water supply for drinking and irrigation'. Additionally, the Spanish Crown insisted that American cities be constructed in a uniform way. Clear-cut, tested techniques – like those associated with building irrigation ditches, missions and urban plazas – undoubtedly helped Spain colonise much of the New World, but Spanish blueprints didn't account for individual landscapes or climates. Urban planners approached Texas the same way they approached California, explains Char, even though the landscapes are distinct. And so, San Antonio was built on a floodplain.

Spanish settlers may have ticked the water supply box, but they must have been on peyote when adhering to another Law of the Indies, this one requiring that settlement occur in places with 'good climate', 'clear and benign' skies and 'without excessive heat or cold'. While reasonable people can argue about the limits of acceptable summer temperatures, Texas skies are indisputably mercurial. On average, San Antonio gets 81cm (32in)[10] of rain a year (compared with London's 68cm (27in[11])), but San Antonio's rainfall often arrives in dramatic bursts (one town received a whopping 56cm (22in) in under three hours[12]).

Some of these assaults stem from hurricanes rolling off the Gulf of Mexico; other times, 'northers' originating in the Arctic deliver rain alongside rapid drops in temperature that can occur in mere minutes. Cause aside, San Antonio's topography makes it prone to lethal flash flooding. Severe floods didn't ravish San Antonio's early settlers every year, but they quickly became a thorn in the city's side. In 1819, two years before Texas became part of the newly independent nation of Mexico, a massive flood tumbled through when waters from the San Antonio River and nearby San Pedro Creek collided. At least 16 people were killed and numerous structures – including several buildings on the city's historic Spanish plaza – were destroyed. A century later, more than 50 people died during a devastating storm, and in 1998 flooding killed more than 30 people while causing economic losses of $750 million.[13]

I remember the October 1998 storm vividly. It happened during my final year of high school, making it especially memorable. Several friends lost their homes, and numerous horses perished at the stables where I learned to ride. We were out of school for days waiting for floodwaters to subside (during a six-day period, 39.8cm (15.7in) of rain fell on the city, a record that went unbroken until 2002).[14] The baseball fields where my brother and I played team sports were completely underwater. Classmates grabbed their kayaks and paddled through muddied water, taking photos and collecting floating debris. Less than a mile away, a massive dam built in the 1920s held back raging floodwaters that would have, in its absence, spiralled towards downtown.

In the words of city officials, Olmos Dam remains a 'first-level of defense' in flood control and, since its creation nearly 100 years ago, water has never spilled over its sides.[15] But it came at an environmental cost. By damming Olmos Creek, the city altered the river's natural features and flow, which negatively impacted native fauna and flora. Building the 592m (1,941ft) dam also required clearing a beautiful stretch of oak forest – an area described in 1849 as 'one of the most beautiful, if not the most beautiful, places in Texas, its woodland grace and park-like beauty so heightened by the perpetual mystery of its profound and noble springs'.[16] More of it disappeared as the surrounding neighbourhood, Olmos Park, came into being. Although the upscale area retains some original oak trees, it was 'totally transformed into an urban landscape that bears little resemblance to its original setting'. Ironically, the low-lying land behind the dam can't be safely developed, which explains why it's a great place to look for owls and other hard-to-spot birds. Trails added in 2014 as part of a city-wide greenway initiative help nature lovers like me explore the flood-torn woods ... when they're dry.

The concrete dam located a short drive from my childhood home in Terrell Hills is San Antonio's most visible flood-mitigator, but other measures – some subtle, some overt – were implemented following the dam's 1926 completion. Streams were straightened to 'convey' (control) the water's flow, and a bypass channel was built around a large bend in the river to insulate the area against flooding. Additional dams were added in the early 1940s as the city installed underground drains and floodgates; and, starting in the 1950s, the US Army Corps of Engineers channelised and straightened almost 50km (31 miles) of the river and its tributaries as part of the San Antonio Channel Improvement Project. The San Antonio River Authority (SARA), was appointed as the Corps' local sponsor. As engineers and urban planners forced the river to comply with their need for a benign urban waterway, the river lost its autonomy, its wildness.

'It was all taken away,' remarks Frates Seeligson, who leads the San Antonio River Foundation (SARA's fundraising arm). 'Urban planners historically saw rivers and creeks as the bane of their existence, so they straightened them and hid them away by putting them in culverts.' This didn't just happen to the river, underscores Frates. Many of the city's creeks suffered the same fate. Nature was lost in the name of flood control and the city's natural beauty suffered as well. 'Although the channelization effectively managed the flood water,' wrote SARA's former CEO, Suzanne Scott, 'it damaged the river's ecosystem and made the river an unappealing drainage ditch.'[17] By contrast, the architects of the Olmos Dam prioritised aesthetics and, as we'll soon see, in downtown San Antonio, beautification and flood mitigation went hand in hand. What happened on the Southside reflects a disparity in resource allocation that has long plagued San Antonio.

While engineers reshaped the city's water flow in the 1920s, a local architect began devising what became the River Walk – a charming series of bridges and walkways located in the heart of downtown. The 'Venice of Texas' was born. To make the pedestrian area seem more natural, SARA planted 12,000 shrubs and trees alongside 5,182m (17,000ft) of concrete and stone sidewalks as part of the San Antonio River Beautification Project (beautiful, to them, must not have been synonymous with natural).[18] Soon, bars, restaurants and hotels opened as formerly wary investors poured money into downtown knowing, or at least believing, that Olmos Dam made it flood-proof. The first restaurant, Casa Rio, opened its doors in 1946; by the time I came along in the early 1980s, the River Walk had become so touristy, local families only ventured to the area for special occasions.

We let tourists have the River Walk and gave them the Alamo, too. Just as the Spanish took San Antonio's natural resources for granted, we overlooked the historic structures built by early Europeans. I'm not Catholic, nor of Spanish descent, so I can see why the missions never spoke to me, but my relationship with the river – or lack thereof – is more complicated. In part, the detachment I felt stemmed from a knowledge gap. I learned about the spring-fed river in school and possessed a basic understanding of the Edwards Aquifer whose artesian zone sits beneath

San Antonio, but I didn't realise that the main spring feeding the river was located a mile away from my childhood home. A former fountain spring, the Blue Hole dazzled renowned landscape architect Frederick Law Olmsted of Central Park fame when he visited San Antonio in 1856. He called the springs 'the first water among the gems of the natural world', writing, 'the whole river gushes up in one sparkling burst from the earth ... The effect is overpowering. It is beyond your possible conceptions of a spring. You cannot believe your eyes, and almost shrink from sudden metamorphosis by invaded nymphdom'.[19]

I would have cycled to the Blue Hole every day after school had it offered anything approaching nymphdom, but the springs that seduced Olmsted lost much of their flow when the city began drilling artesian wells into the aquifer in 1889. According to environmental scientist Gregg Eckhardt, spring flow became 'intermittent and meagre' in the decades that followed as the city withdrew more water from the Edwards Aquifer. By the early 1940s, the springs fell dormant and remained that way until 1973, when months of rain revived San Antonio's 'sleeping beauty'. Heavy rainfall events occasionally awaken the Blue Hole, but even then, the scene is a far cry from the one that bewitched Olmsted.

The River Walk, meanwhile, offered a direct link to the river, but its artificial quality kept me away. Unless a special event lured me downtown, I had no reason to visit. I didn't need or connect with the river or its famed walk, which had become a tourism machine, a gimmick. In the absence of radical change, it was destined to remain a tourist trap.

In the 1960s, environmental awareness swept across the United States as conservationists like Rachel Carson (author of *Silent Spring*) took the podium. Signs of this shift appeared in San Antonio in 1961 when SARA added environmental protection – from pollution prevention to reforestation – to its remit. SARA may have expanded its scope, but conservation wouldn't become a priority for another 40 years.

In the meantime, SARA and the US Army Corps of Engineers continued engineering flood control solutions. For instance, a 4,937-m (16,200-ft) long flood-diversion tunnel located downtown was added in 1997. Critics, meanwhile, pointed out that, in at least some respects, the Corps project was a flood mitigation failure. 'When you straighten a river, the water's velocity increases,' explains architect and urban designer Irby Hightower. Urbanisation increases flow, as well, since impermeable surfaces (like concrete) do not absorb rainwater. As a result, the drainage ditch placed over a southern stretch of the river eroded very quickly. According to Irby, the Corps soon recognised that the channel was unstable, but rather than removing the trapezoidal ditch, engineers proposed widening it while placing concrete on the bottom of the river. The city and county developed similar proposals, says Irby, but grassroots organisations and several local politicians pushed for a new model, one that synchronised environmental conservation, flood control and resource allocation. In 1998, their efforts culminated in the creation of an oversight committee whose diverse members developed new mechanisms for restoring the river (and the cultural connections it used to foster) while maintaining flood control. However, the door to rewilding didn't fully open until 2001, when Congress gave SARA the authority to add environmental restoration to its flood control programme.

Let's chew on this for a second. Flood control that achieves environmental restoration? It sounds paradoxical given the many ways flood control damages the environment, but if any metropolis could pull off the feat, it was San Antonio. The city, you see, had become a world leader in fusing flood mitigation with urban design. Delegations from all over the world came to the Alamo City to learn from the River Walk. More would visit in the coming decades as SARA and the US Army Corps of Engineers began restoring riverine features and functions as part of the $384 million San Antonio River Improvements Project (SARIP).

Twelve years ago, I visited the newly designated 'Mission Reach' of the San Antonio River for the first time. Not yet complete, the 13-km (8-mile) stretch of urban waterway was in the midst of a facelift as part of the Mission Reach Ecosystem Restoration and Recreation project. Part environmental, part recreational, the overhaul sought to restore the river while reconnecting downtown to the historic Missions established during the eighteenth century. Notebook and camera in hand, I was in reporter mode as I conducted research for an article on San Antonio's ecological restoration. My childhood home was only 8km (5 miles) away, but I felt like I was a hundred miles from Terrell Hills and the Olmos Dam. Even the nearby River Walk felt distant, and The Dominion may as well have been on Jupiter.

After pausing to photograph a cormorant drying its wings on a craggy rock, I turned around to catch a view of downtown. I could see the city's tallest building, the Tower of the Americas, and several residential high-rises were in view, but as I spun back, I was struck by how undeveloped, how quiet, the Mission Reach felt. I closed my eyes and tried to imagine the river as it was during Spanish rule.

Then I conjured an even more satisfying picture – one of the San Antonio River before Europeans arrived, when bison and elk came to its banks to drink alongside indigenous people who sustainably utilised its cool waters. A third image of Columbian mammoths and sabre-toothed tigers floated through my mind.

Surrounded by concrete, I struggled to hold on to these images of idyll, but I was buoyed by the fact that the Mission Reach was in the midst of becoming wild again. It will never return to its Pleistocene past, but active rewilding is repairing a broken ecosystem and bringing back locally extinct species.

We already know that the channelisation that occurred between the 1920s and 60s altered the river's flow and decimated its natural beauty. Native plants and trees were stripped away from the banks and replaced by concrete; and natural river features – pools, riffles and runs – were destroyed. In turn, wildlife suffered or disappeared. Guadalupe bass vanished, as did freshwater mussels. Both species indicate ecosystem health, making them canaries in the proverbial coal mine. On a trophic level, they are important players in a robust food chain.

Guadalupe bass are highly intolerant. They only survive when water quality is high and appear to require certain vegetation types, at least in the San Antonio River context, says aquatic biologist Shaun Donovan, who leads SARA's environmental science programme. Freshwater mussels are a different story. They are 'foundational for the food web', explains Shaun, and provide critical cleaning services to the waters they inhabit. Adult mussels can filter up to 45.5 litres (10 gallons) of water a day, making them what several scientists call the 'unsung heroes of our lakes, rivers and streams'.[20] As other experts point out, these unsung heroes simultaneously act as microhabitat engineers.

They alter ecosystems in more ways than one: small organisms like algae live on the exteriors of their shells, making them a bit like hotels that serve bountiful buffets. The nutrients they excrete fuel fellow bottom-feeders. Equally important, mussels help stabilise river bottoms because they anchor into the substrate, which is particularly important during flooding episodes. Together, these functions make freshwater mussels keystone species. Because they exert above-average influence on animal and plant communities, keystone species are paramount to ecological health. Needless to say, the inverse is also true: when keystone species vanish, other species disappear and the environment itself degrades.

Mussels can deliver mighty returns, but they are delicate. Unable to move more than short distances on their own, mussels can't up and leave when environmental conditions deteriorate, and they must live alongside fish to breed (obligate parasites, they attach to host fish to sow their seed, so if fish die out, they do as well). Dams are their downfall. Dams block the movement of fish, thus disrupting the mussel's ability to reproduce, and dams can cause water temperatures to rise. Both impacts put mussels at risk. Across North America, 72 per cent of freshwater mussel species are imperilled, and 30 species went extinct during the past 100 years, with damming of rivers a huge part of the problem. Unsurprisingly, San Antonio's living filters vanished as human engineers developed mechanisms for protecting citizens from flash floods. It's worth noting, however, that the sensitive creatures may have become locally extinct even in the absence of dams: 30–40 years ago, the quality of the urban river was so poor, there were stretches where waste run-off turned the water black.

Shaun knew nothing about freshwater mussels when he joined SARA in 2012. His interest in wildlife began in the marine world and centred on large mammals – the 'charismatic megafauna' that dominate stories about the natural world. Now, 10 years into working for the River Authority, he's a freshwater wildlife convert who understands the importance and vulnerability of freshwater systems. A San Antonio native like me, he is personally, as well as professionally, invested in rewilding our home town. When Shaun talks about it, his professional veneer starts to crack, revealing unabashed enthusiasm. His guard drops when he tells me about his experience encountering freshwater mussels for the first time, 'You're digging around in the substrate and feel what you assume is a rock. You pick it up and see that it's a living animal.' The connection, he emphasises, is 'visceral'.

Shaun recognises the power of tangible connections to nature and the childlike wonder they reactivate. Whenever Shaun brings SARA's executive team to a study site containing freshwater mussels, he invites his bosses to touch and feel the shelled animals. 'Execs have always supported the mussel reintroduction project,' explains Shaun, 'but interactive experiences make them that much more supportive.' Without realising it, the biologist is advocating for the kind of 'personal rewilding' we learn about in Chapter Eleven.

That mussels are the polar opposite of dolphins, lions or pandas is not lost on Shaun. In fact, the mussel's lack of popularity fuels his passion. Shaun went so far as to thank me for writing a chapter on these lesser-known keystone species. Shaun isn't the only conservationist fighting for an inclusive cast of characters and more nuanced nature narratives. Just up the road in Austin, conservationist Merlin Tuttle tirelessly campaigns on behalf of bats. Using a range of techniques, from campaigns to close-up wildlife photography, he helps people understand that bats are interesting, even endearing, and contrary to popular opinion, they rarely endanger humans. His efforts helped Austinites learn to live alongside bats and benefit from them, too (bat-watching tourism generates millions of dollars for Austin each year).

In light of their impressive camouflage and limited motion, mussels are unlikely to bring in the big bucks. Nor are they and Guadalupe bass suitable mascots for the rewilding of the San Antonio River, but if people learn about their ecological *functions*, they might appreciate the rewilding they can't see.

Biologists like Shaun were chomping at the bit to reintroduce wildlife to the Mission Reach, but engineers and scientists had to restore the area first – and before that could happen, the city had to agree on a plan that maintained flood conveyance *and* honoured the priorities identified by the oversight committee. Irby Hightower co-led it. When he and I spoke by phone, I asked him to break down the competing plans for me, learning that there were three options. I call Plan A 'More Concrete'. More Concrete was central to the plans developed by the Corps, city and county, and it probably would have gone forward had Irby and his ragtag team not entered the fray.

Plan B – let's call it 'Nature First' – is my personal favourite. In Nature First, the city would have removed all of the concrete poured by the Corps, torn down the houses and businesses located on the Mission Reach, removed the non-native vegetation growing beside them, and replanted native grasses, shrubs and trees. With Nature First, the clock would have been turned back as the river regained full control, floods and all. But Nature First was never a viable option, not really.

For starters, too much time would have had to pass before newly planted vegetation offered flood mitigation. In the meantime, floods would have endangered San Antonians to an unacceptable degree. Nature First was also too expensive: it would have cost the city at least $600 million, says Irby, when the proposed budget was around $75 million. Not only that, but it would have required some citizens living along the river to sell their homes to the city. According to SARA's communications director, Steven Schauer, critics would have likened voluntary buy-backs of such properties to forced resettlement (more on this thorny issue in Chapter Nine). Conservation has historically separated people from nature, with nature taking priority, but human rights are increasingly incorporated into conservation planning, as they should be. It is worth noting, however, that the city of San Antonio does purchase flood-prone homes from willing buyers in a system praised by Char Miller (cities like Houston have begun to follow suit, and Char hopes California will implement a similar system for residences located in wildfire zones).

The third plan was a compromise, and it prevailed. In Irby's words, Plan C, which became the San Antonio River Improvements Project (SARIP), allows 'as much ecological restoration as can be achieved within certain limits'. Before undertaking it, scientists had to work out what had been lost. By evaluating the Mission Reach's environmental health, they were able to generate a baseline and identify next steps. An assessment conducted in 2004 revealed that, on a scale of 1–10, with 1 representing severe degradation and 10 representing a pristine environment, the Mission Reach stood at a paltry 1.2. Improving that rating will require a multi-pronged approach and heaps of patience (the same report predicted the southern stretch of river would need 50 years to reach 8.2).

Rehabilitation of the Mission Reach started in 2008 and wrapped up in 2014, but restoration was well on its way when I visited in early 2010. Working with the Corps,

SARA began un-straightening the river while using 'design solutions' to replicate natural features, namely three river components that typically follow one another. Riffles (shallow areas containing rocks where water travels quickly) lead to runs (portions with 'average depth and velocity'), and runs lead to pools marked by greater depth and slow flow.[21] Now would be the time to introduce fluvial geomorphology – the study of river shapes, flow and so on – but I'm allergic to jargon and suspect you are, too. Let's leap, then, to the bottom line: before reintroducing wildlife, ecologists had to restore the river and the various habitats found in and around it.

Re-establishing riffles, runs and pools was part of the process, as was a massive replanting effort that spanned 142 hectares (350 acres) of riparian habitat. Grasses and wild flowers were seeded, as were more than 20,000 native shrubs and trees. Together, these changes dramatically altered the district's aesthetic qualities, but the Mission Reach's new plant communities also have critical ecological roles to play. Vegetation planted along the river's banks serves several purposes: it supports wildlife, of course, and filters pollutants and sediments that creep into the river during storm run-off. Plants with deep roots also reduce erosion.

Working with experts from Austin's Lady Bird Johnson Wildflower Center, Mission Reach ecologists used historic records when selecting seeds to scatter. They also made a point of choosing an array of native species to foster a diverse animal assembly that includes full-time residents and passers-by. During migration season, San Antonio is brimming with wildlife since it's located in the middle of a migratory route – the Central Flyway – used by birds and monarch butterflies. I could now make a cringey aside about the river's growing inclusivity (River Walk 1.0 only catered to tourists whereas the new iteration attracts locals, both animal and human), but I'll spare you.

In the coming years, biologists like Shaun closely monitored wildlife along the Mission Reach as they sought to answer a fundamental question: how does ecological restoration benefit animals? We already heard about species requiring reintroduction, but what about the ones capable of coming back on their own? Would they recolonise the area, demonstrating its improving health, or would they stay away, instead choosing greener pastures?

To answer these questions, SARA conducted several scientific studies. One, the sexily named Intensive Nekton Survey, focused on fish and other swimming creatures (nekton stems from the Greek word, *nēkton*, which means 'a swimming thing'). During 2019, the first year of the survey, the team recorded 20 species. The number jumped to 27 the following year. One of the 2020 additions (the grey redhorse) had never been collected in the Mission Reach. Nekton lovers were thrilled. Meanwhile, Guadalupe bass started reclaiming the upper reaches of the San Antonio River with assistance from SARA and the Texas Parks and Wildlife Department. Starting in 2013, aquatic biologists released just under 84,000 of the fragile fish into the River. Most came from a fishery; others were rescued from a nearby river when drought sucked it dry.

Biologists anxiously monitored the new San Antonio River population, fearing that the high-maintenance species might not survive, but they needn't have worried. In the years that followed, juveniles were counted during every sample season, demonstrating that Guadalupe bass were spawning at least once a year. Genetic studies quelled another fear – that the species would hybridise with an invasive bass introduced by anglers. Shaun says it's too early to observe enhanced trophic interactions linked to the reintroduction of Guadalupe bass, but in his opinion, the fish's success proves the river has reached an ecological equilibrium. Trophic interactions, he emphasises, are one measure of that equilibrium.

A second survey, this one on birds, delivered equally encouraging results. Initiated in 2015 – one year after the Mission Reach's official completion – SARA worked with an outside expert, Martin Reid, to count the number of individuals and species using the Mission Reach's restored habitats. In total, 201 species – including a few rarities – were observed (the number has since grown to 206). Martin estimates that no more than 50 bird species used the area before its restoration, although it's impossible to say given a paucity of pre-restoration data. One of the most exciting moments occurred on 16 May 2018 when Martin heard the unmistakable call of a black-capped vireo – an endangered species that isn't known to breed within the city limits. Martin scrambled to photograph it and was ecstatic when he encountered it months later with two healthy fledglings.

An even rarer bird – the stunning blue bunting – caught Martin's eye in February 2021. Blue buntings rarely venture north of Mexico and had never been recorded in Central Texas, so local birdwatchers flocked (pun intended) to the Mission Reach. I was one of them. During the pandemic, I spent months in San Antonio and became a fully fledged twitcher: I downloaded bird ID apps, signed up to birdwatching email alerts and travelled in search of rare sightings, as I did after reading about Martin's blue bunting. Sadly, I never spotted the iridescent passerine, but I loved seeing wildlife enthusiasts gathering beside the banks of the river. Years earlier, the sight would have been as rare as the bunting itself.

Of all the species enjoying the Mission Reach, one stands out. It isn't endangered, nor hard-to-spot, nor an exciting newcomer. Boring, perhaps, but critical to the city's vision for the Mission Reach. I'm talking, of course, about *Homo sapiens*. Getting people to enjoy the area again didn't require a reintroduction per se, but our species didn't passively recolonise the area, either. Trails, parks and public access points had to be created before we could (legally) enjoy the Mission Reach. Put another way, our ability to utilise the restored waterway was never guaranteed.

It's easy to imagine a world in which the Mission Reach was restored behind closed doors or, in this case, fences: plenty of protected areas prohibit or limit human entry, and the Army Corps of Engineers and Oversight Committee occasionally butted heads when it came to the human dimension of the river's rewilding. The Corps worried that human activity was incompatible with ecological restoration, but in the end, San Antonio opted to combine recreation and restoration. Hiking and cycling trails were added, as were kayak chutes, and several public parks have sprung up. One, Confluence, contains sleek pavilions that double as water catchment systems. The net effect is that a once-forgotten part of the city has become a vibrant urban greenway where people can get fresh air and reconnect with nature. People may be at the centre of the Mission Reach, but environmental restoration always trumps recreation, underscores SARA's Steven Schauer. Before recreational features are added, SARA conducts cost-benefit analyses to evaluate possible environmental impacts. Locals aren't always happy with the end result. For instance, they have repeatedly pleaded for lights along the area's trails so they can safely walk and cycle when it's dark outside, but lights have never been added since they would disrupt nocturnal critters. On the other hand, SARA agreed to add a small number of viewing corridors along riverine paths when locals complained that tall grasses and shrubs impeded riverine vistas. However, before taking any action, SARA spoke to bird experts to ensure viewing points wouldn't damage critical bird habitat.

Another kind of corridor can be found on the Mission Reach: 'mission portals' that allow people to easily travel between river trails and the city's historic Missions, four of which are located on the Mission Reach (the fifth and most famous, the Alamo, is in the heart of downtown). Until recently, you had to drive between the Missions – now you can cycle or walk. In part, the city added cultural portals for economic reasons. San Antonio is the poorest major metropolitan area in the country, and the neighbourhood that houses the Missions and eponymous stretch of river is one of the city's most impoverished locales.

For decades, tourism has fed San Antonio's economy. Around 39 million people visit each year; 31 million stay overnight; and the hospitality industry annually contributes just under $14 billion to the local economy. The Alamo and River Walk have long been San Antonio's primary tourist attractions, and experts believe the now-expanded River Walk will lead to further economic gains. It's already happening north of downtown along the aptly named 'Museum Reach'. There, a luxury hotel, culinary institute and retail spaces built in a converted brewery have joined established museums to create an entirely new district: The Pearl. The Pearl's diverse offerings have pumped life into a neglected stretch of San Antonio while bringing significant investment to the city.

Unlike its midtown counterpart, the Mission Reach was never intended to become a bastion of commercial activity, but city leaders hope trails and mission portals increase 'cultural tourism'. In turn, local businesses and communities stand to reap economic benefits. Exponential growth may not be on the horizon – surprisingly, the number of people visiting the Missions has not swelled since the Mission Reach opened – but, if nothing else, their profile is rising. In 2015, the Spanish complexes became the state's only UNESCO World Heritage Site (they're also a national park). The UN agency recognises their 'outstanding universal value' as well as the fact that they represent an 'interweaving of the cultures of the Spanish and the Coahuiltecan and other indigenous peoples'.[22] Interestingly, UNESCO also points to the innovative irrigation ditches (*acequias*) when describing the Missions' cultural value, 'The substantial remains of the water distribution systems are yet another expression of this interchange between indigenous peoples, missionaries, and colonisers that contributed to a fundamental and permanent change in the cultures and values of those involved'.

Drawing attention to those changes makes sense given that San Antonio wouldn't exist without them, but what about their impact on the river, the force that drew people to the area in the first place? The river was San Antonio's lifeblood – at least for a time – and its natural rhythms were disturbed by the very adjustments UNESCO celebrates. When I look at the Mission Reach through this lens – versus, say, an economic one – I can't help feeling a bit uncomfortable.* Conservation and preservation are cut

---

* Not only that, but UNESCO put a positive spin on San Antonio's early history when describing a rich 'interchange' between indigenous Americans and Spanish colonisers. In reality, many indigenous traditions were lost when Spanish missionaries encouraged (and coerced) the region's existing inhabitants to assimilate. On the irrigation front, you will recall that indigenous groups like the Payaya were not farmers that manipulated natural water flow. They only began building *acequias*, a technology introduced from Europe, when Spain established the city's first permanent settlement, one reliant on agriculture.

from the same cloth, and the disciplines have a shared history, with the IUCN born out of UNESCO, but in the context of the San Antonio River, their interests appear to conflict.

I spent days trying to reconcile the Mission Reach's joint aims before realising that, in practice, the project's conflicting baselines are irrelevant. The river's restoration (and the rewilding it facilitated) would never have occurred were it not for cultural, economic and recreational considerations. Some rewilding projects are able to pursue a purer form of restoration since they are financed by wealthy donors like Greg Carr of Chapter One – but, even then, conservationists have to consider how their work impacts people. After all, humans are part of nature, not separate from it, and nowhere is this more true than in cities.

As a child in San Antonio, all of my outdoor activities occurred in backyards, botanical gardens and zoos, or on school playgrounds. It's very American, this habit of playing on private property, but you're left with little choice in Texas given that about 95 per cent of the state is privately owned. Although San Antonio has several public parks, they were considered unsafe or were in disrepair by the time I came along. I was lucky to have forests and farms to roam, but what about children living in urban apartments or houses with small backyards? How were they supposed to get their nature fix? The myriad benefits of spending time in nature have been documented by experts like Richard Louv (author of *Last Child in the Woods*), but acknowledging that we suffer from 'nature-deficit disorder' is just one step in the right direction. To rewild ourselves, we need wild places.

When the founders of the rewilding movement put pen to paper, they emphasised large core areas, but conservationists have since recognised that rewilding can occur almost anywhere, on almost any scale. Which is a good thing, in my opinion, given the urgent need for environmental action and the importance of inclusivity in conservation. For the sake of people, and the planet, conservation and its benefits must be shared. In urban areas like San Antonio, rewilding projects are constrained by urban development, densely packed populations, and other forces that don't inhibit the rewilding of large and remote wilderness areas. However, urban rewilding has a key advantage: accessibility. Half of us live in cities now, and the UN predicts that this figure will climb to 68 per cent by 2050. Urban rewilding can reach (and benefit) more people than remotely located restoration projects.

Ten years after my first visit to the Mission Reach, I returned to the spot where I photographed a cormorant drying its wings. This time, I was in visitor mode. It was April 2020, and I had very little work on my plate, having recently submitted my proposal for this book. While waiting to hear from Bloomsbury, I focused on improving my bird photography skills, and the Mission Reach was the perfect place to hone my craft. Late one afternoon, I made my way to a part of the river I'd never visited before. Beneath a large bridge were dozens of cave swallow nests, and I was determined to photograph parents feeding their chicks as golden light illuminated the water below. Standing on a path beneath the bridge, I watched as adult swallows swirled through the air acrobatically hunting insects. Over and over, I attempted to photograph them capturing meals mid-air, but it was no use: I couldn't anticipate their flight paths, making in-focus hunting shots impossible. Putting my camera away, I looked up the path and scanned the banks of the river. On one side, a family gathered for an outdoor picnic. On the other, a group of cyclists whizzed by. It was a simple and timeless scene of the kind captured by impressionist painters. Another image of idyll, this one completely different from the three I conjured a decade earlier. As I stood there taking it all in, I momentarily forgot the city's complex relationship with the San Antonio River and allowed myself to revel in rewilding's afterglow.

CHAPTER SIX

A Conservation Compromise in Post-War Rwanda

On 6 January 2020, I travelled to Rwanda for the first time. I'd wanted to visit the 'Land of a Thousand Hills' for years and was eager to explore Akagera National Park, which is located on the eastern edge of the small landlocked nation. With lakes and papyrus swamps dotting its core, Akagera features spectacular landscapes and contains the largest protected wetland in Central Africa, giving it critical conservation importance. The park has enjoyed protection for just under 90 years, but its very existence came under attack during the 1990s when Rwanda endured an unspeakably dark period that culminated in the 1994 Rwandan genocide.

Although peace brought the populace some measure of justice and stability, Akagera's woes escalated during Rwanda's post-war period. At times, its troubles seemed insurmountable. Then, in 2009, an ambitious rewilding project began when the Rwanda Development Board (RDB) signed an agreement with NGO African Parks (APN) to manage and restore the national park. Although many milestones, including species reintroductions, have since been reached, Akagera remains relatively unknown to foreigners like me, which is one reason I chose to visit in early 2020: I wanted to experience the park during its regrowth to witness the rewilding process, growing pains and all.

My time was limited, so when my friend Laura (a fellow American journalist) and I ironed out the logistics of our trip, I suggested heading straight to Akagera after landing in Rwanda, but Laura had a different idea. She hoped to spend our first morning at the Kigali Genocide Memorial. At first, I resisted her suggestion. Having just read several books about the genocide, I didn't feel like I had the fortitude to tour the site in a few short hours before switching gears. And even though I knew that Akagera's history was inseparable from the atrocities of 1994, my instinct was to compartmentalise the two. This trip is about wildlife, I naively thought to myself. I didn't have the time or emotional bandwidth for anything else.

In the end, we agreed to spend the morning at the memorial. Our driver and guide, Theo, waited in the parking lot while Laura and I silently walked through exhibits that brought to life the horror I had read about in *A Sunday at the Pool in Kigali*. A video segment on repeat in one of the museum's rooms showed young men and women talking about the genocide with incredible candour. I was astonished to see their vulnerability on display, but soon learned that such openness is common in Rwanda (Theo, for instance, shared a deeply personal story about barely escaping death only to learn that his father had been murdered). A video segment on repeat in one of the museum's rooms showed young men and women talking about the genocide with incredible candour. I was astonished to see their vulnerability on display, but soon learned that such openness is common in Rwanda.

This insistence on directly confronting trauma for the sake of healing broke apart my instinctive compartmentalisation. More importantly, it reminded me that you can't separate the country's identity from the turmoil it experienced during the second half of the twentieth century. In the same way, Akagera's story is intertwined with the country's civil war, genocide and eventual peace process. Even in a book on rewilding, viewing the park's ecology and wildlife in a vacuum isn't just inappropriate – it's impossible.

Gaël R. Vande weghe and I were born in the same year: 1981. We both loved animals, nature and exploring, but while I spent my childhood tromping around my family's Texas ranch, Gaël ran wild in Akagera, where he and his family lived from 1985 to 1989. Gaël's Belgian father moved the family to the national park while coordinating an ecological assessment of its fauna and flora. The survey was intended to inform a management plan for a landscape that's been protected since 1934. His expertise lay in botany and birds. Having observed Akagera's birds on foot for years, Gaël's father knew the park's avifauna better than most trained biologists – and his passion rubbed off on his son.

Gaël's memories of this enviable period are crystal clear. From a young age, he made his own records, sketching what he saw in worn notebooks and comparing his drawings to field guides. By the age of eight, he knew the park's wildlife well enough to guide day visitors. Sitting in their cars, he pointed out familiar animals, his eyes shining with excitement whenever he came upon his favourite bird species – the red-faced barbet.

When we spoke in autumn 2020, Gaël told me that he tallied every leopard sighting (eventually reaching 42) and laughed as he recalled the safari vehicles then in vogue (minibuses painted like giraffes and zebras). Akagera housed upwards of 320 lions at the time, making sightings within and beyond its unmarked boundaries frequent. 'Lion encounters stay in your memory forever,' says Gaël, putting into words a sentiment I and countless others feel. Tourism thrived during the second half of the 1980s, and to the fledgling naturalist, life seemed uncomplicated.

Gaël had superficial knowledge of the conflict between Rwanda's largest groups – Hutu and Tutsi – since some of his Tutsi relatives sought refuge in neighbouring and distant countries, from Uganda to Japan, when ethnic tensions boiled over in the late 1950s. At the time, Rwanda started transitioning from a Belgian colony that favoured elite Tutsis to an independent republic dominated by Hutu leadership. During this phase, now called the Rwandan Revolution or Wind of Destruction, more than 330,000 Tutsis fled the country.

Gaël's parents met in Burundi and got married in Belgium – where Gaël was born. When he was an infant, the family moved to Rwanda. Like many protective parents, Gaël's shielded him and his sister from escalating tensions associated with the impending return of exiled Tutsis who, under the leadership of Rwanda's current president Paul Kagame, formed the Rwandan Patriotic Front (RPF) from exile in Uganda. Gaël's parents' instincts were partly inspired by a desire to maintain their children's innocence for as long as possible, but they also worried about their children telling other kids about their Tutsi heritage without recognising the risks involved.

In 1989, Gaël's family moved to Kigali so he and his sister could attend school, but weekends and vacations were frequently spent in Akagera. During the last weekend of September 1990, Gaël and several friends who had started a birding club travelled to the national park for a mini-safari. Their weekend was idyllic – a montage of simple pleasures, young friendship and reflective moments in nature – but chaos waited around the corner.

On the afternoon of 1 October 1990, rebel RPF forces entered Rwanda from Uganda and declared war on the Hutu-led government. Surprise attacks had occurred before, but a 1992 Human Rights Watch report said the October strike was the first time Rwandan exiles 'posed a serious threat to the government'.[1] It marked the start of the Rwandan Civil War, a conflict that only came to an end in July 1994 following a four-month genocide in which 800,000–1,000,000 Rwandans, most of them Tutsi, were killed.

A few days after the RPF attack in the north, rebel forces made their way towards Kigali. The president saw this advance as an opportunity: he orchestrated a fake attack on the city, claimed it was led by RPF forces and used the alleged incident to justify the incarceration of thousands of Tutsis (according to Human Rights Watch, some 13,000 were imprisoned and dozens were killed). When Gaël's mother learned that her name was on an arrest list, she and the children rushed to the airport where Belgian troops loaded them onto a plane bound for Europe. Gaël's father remained behind with hopes of salvaging the research materials he stored at the family's Akagera home, but he never made it inside the park – someone waved him down on its outskirts to alert him to the fact that several park workers had been killed. He left the country soon after to meet his wife and kids in Belgium. The Vande weghes stayed in Europe for a year before relocating to Burundi. There, they waited for the war – and later the genocide – to end.*

The family travelled home a week after peace was established in July 1994. Gaël felt safe but found himself wandering through a world of unimaginable loss. 'The country was like a cemetery of dead bodies,' he tells me. 'There was nothing left – no water, electricity, or doors on

---

* What took place between 7 April and 15 July 1994 is nearly impossible to put into words. As historian Saul Friedlander stated, genocide is a history 'too massive to be forgotten, and too repellent to be integrated into the normal narrative of memory'. (Friedländer S. 1993. *Memory, History and the Extermination of the Jews in Europe.* Indiana University Press, Bloomington, USA.)

houses.' Nevertheless, he and his father drove to Akagera within a matter of days to check its status.

Because the park was laden with landmines, they weren't able to enter, so they drove along the main road that flanks Akagera's western boundary, scanning in every direction as they headed north. Gaël spotted numerous antelopes and zebras, but other large mammals – elephants and lions in particular – were noticeably absent.

Unbeknownst to Gaël, the lions he revered as a boy continued to roam Akagera's grasslands in large numbers, but three years later, in 1999, a single event wiped them out, making lions extinct in the region – and Rwanda.

On the morning of 29 June 2015, Simon Naylor and his conservation colleagues at the Phinda Private Game Reserve featured in Chapter Four gingerly loaded five sedated lionesses into crates bound for Akagera. Several charter flights later, the conservationists' counterparts at African Parks opened the cats' crate doors, sending the fivesome into a 1,000m^2 (10,764 square foot) enclosure, where they joined two males translocated from another South African reserve. It was the first time lions set foot in Akagera – or Rwanda – in 16 years, a milestone widely celebrated in national and international media outlets. Following a 14-day acclimatisation and quarantine period, the lions left their temporary home and began wandering north to form prides of their own.

It didn't take long for the next generation to appear: 11 playful cubs were born in the spring of 2016. Just like that, Akagera's population doubled. The tawny felines had ample land to roam and prey to hunt, but unlike the lions of Gorongosa – who were for many years the only carnivores in a massive, prey-packed zone – Akagera's lions had to compete against existing predators, hyenas and leopards for food. With plenty of water sources and food to go around, Akagera's pride thrived, but healthy long-term growth requires genetic diversity. Lions aren't found in any other parts of Rwanda and don't migrate into Akagera from Tanzania (which borders the park's eastern boundary), so in 2017 African Parks introduced two additional males, sending them into the park with fingers crossed for a second pride and further cubs. Such hopes were realised – and exceeded. At the time of writing, 40 lions inhabit Akagera's acreage.

Lions are prolific breeders, and their position at the top of the food chain makes growth relatively easy. Collaborative hunting – and parenting – also helps (human parents, take note). Which isn't to say that lions have it easy. Quite the contrary, as wild cat experts from Panthera make clear: lions have disappeared from 95 per cent of their historic range[2] and as few as 20,000 remain in the wild today. Many threats are familiar – habitat fragmentation and human–wildlife conflict are at the top of the list – but in recent years lions have been killed for their body parts and bones as traders seek substitutes for increasingly scarce tiger parts long-consumed in parts of Asia. With similar builds, jaguars and leopards are also poached to fill the void left by a shrinking global tiger population, which has been reduced by 95 per cent in just over a century.[3]

When it came to restoring Akagera's wildlife, lions were the obvious starting point. Historically, they lived in the park in high numbers; they only vanished because of human interference (as we'll soon see); and, like all apex predators, lions play an important ecological role by keeping prey species in check, which in turn prevents vegetation overgrazing. According to African Parks, whose speciality is landscape-wide restoration, Akagera's trophic dynamics needed a boost that only reintroductions could provide. As a 2013 report put it, the return of lions promised to 'introduce a natural force to govern herbivore populations and increase biodiversity' – and by 2013 the conditions were right. Most species were rapidly growing year by year, with eland and roan antelope the only ungulates whose numbers were merely stable.

When I investigated the reintroduction that took place two years after the report's publication, in 2015, I assumed Akagera's wildlife had precipitously declined *during* the civil war. What I discovered surprised me. Although both sides of the conflict used the park (and its natural resources) during the war, Akagera's story is distinct from Gorongosa's. The two share themes but follow different arcs. In one, civil war was the enemy of conservation; in the other, peace paved the way for environmental destruction.

In the words of African Parks, 'While peace was finally restored in the 1990s after the 1994 genocide, Akagera's demise was just beginning.'[4] Some would say it was inevitable, even necessary – Rwanda's only chance at healing. What Akagera offered was arable, unoccupied land, a precious resource in high demand as exiled Tutsis returned home in the wake of the genocide.

No one knows the precise number of refugees who entered the country after years or, in some cases, decades away, but experts estimate that several hundred thousand people and upwards of 700,000 cows flooded into Rwanda in the mid 1990s. Many hoped to pick up where they left off but quickly discovered that their land had been redistributed following their departure, leaving them – and their livestock – homeless. In more spacious parts of the world, displaced individuals could have settled in undeveloped areas, but Rwanda has been severely overpopulated for centuries, making land scarcity a long-standing problem. It wasn't always this way. Before Rwanda was colonised – first by Germany in the late nineteenth century and then Belgium in 1916 – the country utilised a collective land-ownership system overseen by Rwanda's king and several of his chiefs. Colonisation upended this system. According to journalist Philip Gourevitch, 'Ninety-five per cent of Rwanda's land was under cultivation' by the mid 1980s 'and the average family consisted of eight people living as subsistence farmers on less than half an acre'[5].

Anticipating such shortages, the Arusha Peace Accords that put an end to civil war and governed the terms by which refugees returned instructed the Rwandan government to make available all unoccupied land.[6] Akagera's fate was sealed: the national park and a neighbouring hunting reserve were open for resettlement, and in 1997, 60 per cent of the combined area lost protection, its vast grasslands set aside for people and livestock.

When created by the Belgian colonial government in 1934, Akagera occupied 250,000 hectares (617,763 acres), but only 112,000 hectares (276,758 acres) remained after it was degazetted (in conservation parlance, the term refers to a protected area losing status or size). A scientist writing in 1998 decried the park's 'amputation'.[7] He acknowledged that the government planned to preserve Akagera's most biologically rich regions but worried that carving it up would prevent the natural migration of certain species, which would, he said, reduce numbers and potentially cause local extinctions.[8] African Parks calls it the 'most significant, human-induced' event in the park's recent history, noting that the background to the government's decision was complex.[9] Compromise is unavoidable in conservation (as in everything), but critics point out that few reductions of this magnitude have occurred in recent African history. Gaël tells me that his father, who was part of the commission tasked with managing Akagera's resources after the genocide, was disappointed by the deal the government struck: the reduction was, he felt, unnecessarily extreme.

On the other hand, the government found itself between a rock and a hard place. It couldn't adequately support displaced refugees without compromising some of its natural resources. On top of that, Kagame's administration had to consider the wider historic context. When Akagera was created in 1934, people were pushed off their land. Additional restrictions would have been a double blow to families whose land was previously taken away. In the end, much of the park was handed over to people even though its degazettement put the country's natural resources at risk. Another day, another compromise.

On the surface, it may seem like the Rwandan government undervalued the country's natural heritage – the split, after all, wasn't 50/50 – but a well-placed source says otherwise. Mark Rose is the CEO of Fauna & Flora International (FFI), the UK NGO that led Operation Oryx (Chapter Two). FFI has worked in Africa, and on gorilla conservation, for decades. Mark has, too. According to him, when the RPF remained a rebel force, RPF leader and now-president Paul Kagame sent a telegram to the International Gorilla Conservation Programme (IGCP) emphasising that he knew the gorilla's value to the country and would do everything in his power to keep the endangered primate safe during turmoil. 'Kagame fully recognised the importance of mountain gorillas to the wealth and revenue of Rwanda,' Mark told me during an autumn 2020 phone interview. 'He became a great champion of their conservation from the day he took office.'

Months later, after peace arrived, Mark and other conservationists associated with IGCP convened to take stock of Rwanda's gorilla population and discuss how conservation – and wildlife tourism – could safely resume. Akagera was specifically mentioned during the three-day workshop. 'It was rumoured that the incoming government had done a deal with the returning livestock farmers to give them two-thirds of the park,' adds Mark. 'We conservationists tried to push back, but the deal was done.'

Pastoralists and their livestock streamed into the country in the thousands and made their way to Akagera. Together, the park's new inhabitants altered its habitats and threatened its wildlife, but Akagera's hooved newcomers arguably created more problems than their keepers. You see, domesticated herbivores consume any and all plants (and trample many, too) whereas their wild counterparts are selective about cuisine. Overgrazing can prevent future seed germination while fostering the growth of unpalatable (or even toxic) plant species, as occurred in one of Akagera's sectors. As certain plants disappear, soil is prone to erosion.

Farmers evidently started out grazing their cattle in a portion of Akagera that lost its protection in 1997, but as resources dried up, they moved into the park itself. Cattle drink more water than wild ungulates, so the park's herds were led to the park's largest lake, Ihema, along whose banks I stayed during my 2020 trip.

The lake is breathtakingly beautiful, its surfaces covered in floating papyrus plants that I assumed were fixed in place when I admired them during my first evening at Akagera's Ruzizi Tented Lodge. It was only on day two, when Laura and I met for a cold beer on the lake's edge, that I noticed something strange: the mass we noticed the evening before had moved. A day earlier, it was around 30m (98ft) from the deck that overhangs the lake, but it now appeared to be no further than 10m (33ft) from where we stood. We shot each other puzzled glances. Laura could blame jet lag (she flew to Rwanda from New York, whereas I arrived from London), but I couldn't pin my apparent madness on a few sips of beer. Just then, a waiter came out to check on us. Did we need another drink? No, but we needed a third pair of eyes. Fortunately, he quickly solved the papyrus mystery. We weren't crazy: the plants had moved. They do it all the time, their verdant clumps zigzagging across the lake when pushed by the wind.

The scene would have been far less picturesque in the late 1990s when cattle encircled Lake Ihema. Akagera's integrity was under attack, wrote the author of one journal article, 'At this rate and in this perspective, [the park] as it was before the war seems doomed'.[10] Others feared that overgrazing would lead to a proliferation of rodent populations, locust swarms and a spike in tsetse flies. Aerial surveys conducted in the 1990s and 2000s confirmed some of these fears: between 1994 and 2002, the park's wild animal populations shrunk by up to 80 per cent through a combination of hunting and agriculture.

Akagera's vegetation changed as well. Native grasses were lost, and trees and shrubs proliferated, as did invasive plant species, including one that produces toxic chemicals that impede the growth of competing plants. Given the negative impact cattle have on the growth of some plants, how did this invasive species proliferate in the presence of so many domestic grazers? The authors of one study hypothesised that it was more resistant to cattle-induced soil disturbance than many native shrubs.[10]

Some conservationists believe that livestock and wildlife can coexist – Kenya's Ol Pejeta Conservancy moves cattle into predator-proof enclosures at night to stave off what's known as livestock predation – but more often than not, human–wildlife conflict erupts when domesticated and wild animals live side by side. The main cause? Predators like hyenas and leopards attack livestock, which in turn prompts angry farmers to retaliate. Similar issues arise all over the world and account for why some Brits are opposed to proposed reintroductions of lynx and wolves, though in many cases people react to perceived – versus real – threats. Farmers also lose time and money when animals trample or consume their crops, and many locals fear animal encounters – rightly so (however, contrary to popular belief, hippos and crocodiles account for most African wildlife deaths). On the other hand, cows aren't merely a source of income; for pastoralists, they symbolise prestige. 'Faced with this dominant symbolic status,' wrote one scientist, 'wild animals do not represent an important asset in the eyes of people compared to livestock.'[11]

Alphonse Ntabana sees both sides of the coin. The Ugandan-born wildlife guide works at Akagera's newest lodge, Wilderness Safaris-owned Magashi Camp, and grew up in one of the refugee settlements located in the old Akagera National Park. His family fled the country in the late 1950s and returned decades later to find that their land was occupied by total strangers. When Alphonse first set foot in Rwanda – three years after the end of the genocide – he was 13 and had no idea that his family was settling on parkland. Leopards and elephants were prone to wandering into his village, and he occasionally heard lions roaring at dusk, but to Alphonse's community, wildlife was a threat.

'We didn't want them coming to where we were,' he says. 'Most people didn't like animals because they just caused us trouble. For instance, buffalo killed people, and hyenas and leopards sometimes stole our cows. For us, life was just cattle and grazing, and that's the only life people wanted to live.'

Despite this culture of pastoralism, Alphonse developed an interest in wildlife. While attending school in a town 10km (6 miles) from his village, he missed the constant sight of wild animals and regaled his urban classmates with stories about zebras and baboons, smiling as their eyes widened in wonder. For them, zebras and baboons were the stuff of stories and textbooks, but Alphonse had experienced their splendour and mischief first hand. In school, Alphonse fed his interest in the natural world by studying biology and geography, but when he attended university in Uganda, he decided to focus on tourism since it dovetailed with his passion for nature and wildlife and was, as he explains, a 'space with professional opportunity'.

By the time Alphonse graduated in 2013, African Parks and the Rwandan government were in their third year of a 20-year partnership to restore Akagera, but several years would pass before Alphonse considered the national park a professional prospect. It wasn't on his radar – not yet.

During its first few years managing Akagera, African Parks focused on surveying animal populations, overhauling law enforcement and finding ways to keep people (and their livestock) out of the protected area. The latter posed the greatest challenge, and it eventually became clear that the objective wouldn't be met without a physical boundary demarcating the park's perimeter. Even if communities had fully respected the unmarked border between their new homes and the park, animals would have continued passing back and forth – bringing danger to people, cattle and themselves. Wildlife continued frequenting community land despite the fact that degazetted land rapidly deteriorated as locals cut down trees, converted grasslands to agricultural plots and hunted bushmeat. Livestock predation, which occurs around the world, led to retaliation. Herders took to poisoning livestock carcasses to keep lions away; such incidents affect other species, too (a single one can wipe out hundreds of vultures).

Prior to the civil war, 300 lions called Akagera home, and while it's unclear how many survived the war and subsequent resettlement period, a 1999 report said lions were present but very scarce.[12] By the end of the year, the species was exterminated.

The extinction event began when several farmers poisoned a cow carcass and left it overnight with hopes of targeting lions that were, in their eyes, pests. When the men returned to check on the carcass the following day, they discovered pawprints beside four deceased lions and decided to follow the cat that got away. They traced its tracks and soon encountered the lion, but the lion reacted by mauling the men, injuring three before succumbing to spear-inflicted wounds. It wasn't clear until later that the fallen lion was Rwanda's last.

Human–wildlife conflict of this kind occurs all over the world. Elephants eat crops; big cats prey on livestock; coyotes and foxes kill pets. People can't blame animals for acting on their instincts – especially in light of the fact that *we're* the ones modifying the planet in ways that exacerbate conflict – but we can enact mitigation measures. One involves financially compensating livestock owners for their losses; another focuses on separating people from wildlife. Akagera uses both techniques. Like all of the country's national parks, it participates in a national compensation scheme. Five per cent of all tourism revenues go into a fund that compensates communities negatively impacted by human–wildlife conflict. To separate communities and wildlife, Akagera constructed a 120-km (75-mile) long electric fence in 2013. It runs along the park's western boundary (the eastern edge has a natural one, courtesy of the Kagera River). Park manager Jes Grüner oversaw the effort.

After erecting the fence, Jes and his team deliberately collapsed several portions of it and used helicopters to corral animals lingering outside Akagera into the park. Like translocations, the costly manoeuvre underscores that wildlife conservation often requires deliberate interventions. In the context of rewilding, interventions like these can seem paradoxical. Rewilding aims to restore nature to a self-sustaining state, but to get there conservationists often have to kickstart the process. Fencing, a critic might point out, is unnatural. By extension, Akagera is incapable of returning to a genuinely wild state. Wildness, of course, is subjective, making ecological baselines critical. Are conservationists hoping to recreate a pre-human world, or are they undoing the recent yet extensive damage caused by people during the past few centuries?

When I asked Jes about his attitude towards baselines, he articulated a simple philosophy that goes something like this: 'These animals used to thrive here; humans put them at risk; they now need secure habitat; we can provide that; so we should.' He was a purist when he started working at Akagera 12 years ago – he only believed in reintroducing species that were unquestionably present before the civil war. Now, he's not so sure. 'Seeing the state of affairs these days, I'm open to reintroducing species like cheetah and white rhino* even though they haven't been in this area for a very long time.' (Jes can only extrapolate by looking at

---

* In November 2021, months after I completed this chapter, Akagera introduced 30 white rhinos to its acreage in what has been called the largest rhino translocation in history. As with Akagera's lion transplants, the southern whites came from Phinda.

the species' historic range, but new technology that detects trace DNA in soil samples may soon fill in the gaps.) 'The purist approach means losing species,' he adds, though he's quick to point out that theoretically reintroducing cheetahs to Rwanda is poles apart from the Indian government's controversial plan to send African cheetahs to India as ecological replacements for the critically endangered Asiatic cheetah, a subspecies now relegated to Iran and hanging on by a thread.

According to Jes, the fence has exclusively delivered positive results. It stops animals from getting into conflict with nearby communities – which keeps people *and* animals safe – and prevents resource competition between livestock and wildlife. It also gives Akagera's animals, some of which are vulnerable to poaching, an added layer of security. Akagera is not alone in putting up an artificial boundary: national parks across Africa are increasingly fenced by necessity. Besides, as one ecologist points out, all wilderness areas have natural borders like mountain ranges or rivers, so boundaries are not necessarily problematic. Poaching and illegal fishing still occur in Akagera, but alongside improved law enforcement and daily foot patrols, the fence has put a serious dent in illegal activity. Together, these tactics have gone a long way towards helping Akagera's wildlife reach pre-war levels. In theory, many native species could have bounced back on their own, but although some showed signs of organic recovery before the fence arrived, scientists point out that growth occurred at a very slow pace.

After years away, Alphonse returned to the Akagera ecosystem in 2015 when African Parks hired him to work as a community guide. Such guides don't have fixed itineraries: they wait in the visitor centre hoping day visitors turn up asking for a local expert to show them the park.

Alphonse enjoyed the work but quickly discovered that most visitors demand to see the 'Big Five'. It's a common and frustrating demand, both for wildlife guides and travellers like me who prefer an organic safari experience over a bucket-list trip, but it's hard for guides to convince a novice that a lilac-breasted roller (my favourite African bird) is more interesting than a pride of lions (it is, take my word for it). When lions returned in 2015, Alphonse was relieved – the big cat's presence meant guests had a higher chance of going home happy – as well as personally delighted. 'Hearing lion roars at night is a very special thing,' he tells me, 'because the sound was absent for such a long time.'

Jes and his team closely monitored the pride's progress while preparing to reintroduce another iconic species: the black rhino. Of Africa's two – black and white – the black rhino is the most imperilled (the most endangered *subspecies* is the northern white, as we learned in the Introduction). The IUCN considers black rhinos critically endangered, and the species' precipitous decline took place during an alarmingly short period of time. Numbers dwindled by 96 per cent from 1970 to 1992, largely because of hunting and poaching, and at one point only 2,400 individuals remained. Although the black rhino's future is far from secure – approximately 5,500 survive today, and populations are growing – conservationists staved off extinction using two primary approaches: (1) translocating rhinos from vulnerable areas to secure ones like Akagera and the private reserves of Chapter Ten; and (2) criminalising international trade in rhino horn via the Convention on International Trade in Endangered Species of Wild Fauna and Flora (CITES). The latter helps all rhinos but is critical to the black rhino's survival because an estimated 95 per cent of the horn sold in Asia comes from the species.

Their horns make rhinos tremendously vulnerable (despite being composed of the same protein, keratin, found in human hair and fingernails), which in turn makes looking after them an expensive, time-intensive commitment. Not all parks or reserves have the resources to keep high-value herbivores safe. Indeed, some landowners got rid of their rhinos during an African poaching spike that began in 2007 and peaked in 2015. African Parks, meanwhile, was committed to reintroducing black rhinos to Akagera but needed time to ready the park.

Security was of paramount importance. The fence was a critical first step, but African Parks added a helicopter and canine unit to its arsenal before welcoming back rhinos in 2017, and the anti-poaching squad grew to meet Akagera's need for additional security. Special training – delivered at Akagera and international sites that already contained rhinos – helped Akagera's rangers learn how to track the herbivore. Top-of-the-line technology lets the team monitor animals in real time. Of course, sharing such information is risky, so Jes and his colleagues rely on a tool developed by SmartParks that provides long-range communication and is, in Jes's words, 'pretty much unhackable'. It's also solar-powered.

When it came to selecting Akagera's founder population – a group that's capable of repopulating the region on its own – the team chose 18 eastern black rhinos from South Africa's Thaba Tholo Game Ranch (in total, the animals travelled 3,999km (2,485 miles)). 'We started by translocating eight males and ten females, all adults,' says Jes, 'with the knowledge that Akagera can ultimately provide safe harbour to 150 black rhinos.' Eighteen wasn't a magic number, however. Because the park's rhino history is murky, African Parks didn't have a clear pre-war baseline.

Ecological research has been conducted at Akagera since the park's 1934 founding, but certain pictures are clearer than others. Which is not surprising given that some species, including rhinos, are more elusive than highly social, easy-to-spot creatures (e.g. impala, baboons, even lions). Moreover, the park's boundaries and size have changed over time, so Jes and his African Parks colleagues only feel comfortable analysing wildlife trends in the current park from 1997 onwards. However, extrapolations can fill in gaps, as can fragments of information, such as an 1863 statement by explorer John Speke that sheds light on the region's other mega-herbivore: the elephant. Speke wrote about an area in (or very close to) Akagera where 'vast herds of elephants' were 'driven off' by hunters seeking ivory.[13] Speke is famous for having been the first European to reach Lake Victoria, which he correctly identified as the Nile's source; he also spent a great deal of energy promoting the racist Hamitic Hypothesis that distinguished Hutus and Tutsis and undoubtedly contributed to the ethnic conflict that later tore the country apart.

Several other nineteenth-century sources alluded to elephants, but none were seen in Akagera during a 1938 survey. A lone skull and two records from the 1960s suggest that at least several individuals moved through the park during the middle of the twentieth century, but experts believe the species disappeared in the 1960s and only returned in 1975 when 50-odd pachyderms were translocated from southern Rwanda. No one knows how many survived the war, but in 2010 Jes and his team used aerial surveys and tracking collars to assess the park's population, concluding that it numbered between 76 and 80. 'They were scared as hell and keeping to themselves in hard-to-reach portions of the park,' says Jes. 'But now they're relaxed, having plenty of babies, and going to areas they never visited before.' At the time of writing, approximately 130 elephants call Akagera home.

When – and for how long – black rhinos occupied Akagera is less certain. According to a 1969 journal article, they vanished from the park by the late 1950s, their absence no doubt caused by trophy hunting and poaching as well as rinderpest, a cattle plague that swept across Africa at the end of the nineteenth century. Park managers reintroduced black rhinos in 1958, and within 30-odd years, six animals translocated from Tanzania grew to more than 50, but the civil war stripped Akagera of its black rhinos once more. The species was deemed locally extinct in 1994, although a single individual was spotted south of the park in 2007.

Alphonse calls the homecoming that took place a decade later an important milestone in the park's rewilding. He'd never seen rhinos before and, having grown up hearing stories about them chasing community members, was nervous about their presence. His fears weren't unfounded: tragically, one of Akagera's rhinos killed a veteran conservationist and tracker soon after the May 2017 reintroduction. Hungarian Krisztián Gyöngyi had spent more than five years studying and tracking rhinos at two Malawian sites managed by African Parks and was in the midst of obtaining a PhD on the conservation ecology of black rhinos in Malawi's Liwonde National Park. His death is a sobering reminder that conservationists sometimes put themselves in the line of fire.

Fear wasn't on Alphonse's mind when he first laid eyes on the park's newest wild residents. He was exhilarated by their presence. Initially observing the newcomers through the slats of their pre-release bomas, Alphonse spent time studying their features. Many people incorrectly assume that colour distinguishes black and white rhinos when in reality the term 'white' stems from the Africaans word for 'wide' (*weit*). White rhinos have square lips whereas black rhinos' are hooked.

These semi-captive observations prepared Alphonse for safari sightings with guests. Rhinos are extremely elusive, but travellers are desperate to see them, so encounters are always hard-won. 'They really wake you up,' says Alphonse. 'And they give us the chance to talk about the threats facing rhinos, the park's history, and how African Parks and the government brought them back.'

In 2018, two calves were born, bringing the park's population to 21, but African Parks decided to translocate five additional animals, this time from European zoos, in June 2019. Akagera's founder population didn't need the boost, says Jes, but African Parks was happy to welcome new blood when several European zoos teamed up to send healthy captive black rhinos back to Africa as part of the European Association of Zoos and Aquaria (EAZA) *ex situ* programme.

As we saw in Chapter Two, *ex situ*, or captive, breeding efforts focus on maintaining and beefing up healthy populations in accredited facilities. Captive groups are primed to act as 'insurance populations' in the event that the species goes extinct in the wild, but zoo-assisted rewilding doesn't only occur in the face of extinction. Entities like England's Chester Zoo – which leads EAZA's black rhino *ex situ* programme – and the Czech Republic's Safari Park Dvůr Králové are eager to send healthy rhinos to their native habitats when possible. By donating two males and three females to Akagera, they were able to increase the genetic diversity of the park's population while drawing attention to the role zoos play in conserving wildlife.

Four of Akagera's captive transplants came from facilities in Europe, and one travelled from England's Flamingo

Land, but the newcomers spent time getting to know one another at the Safari Park Dvůr Králové in the Czech Republic before making the 6,000-km (3,728-mile) journey to Akagera. Unfortunately, one of the two males sent to Akagera in 2019 died of natural causes eight months after arriving in Rwanda. Jes says he never adapted to African life, 'From day one, he was slightly stressed, slightly lazy and didn't eat as much as the others. We call it "maladaption".' His captive upbringing may have been to blame, but the stress associated with translocation can't be overlooked. Rhinos are prone to stress, explains Jes, making them more vulnerable to such incidents than more adaptable species. There's no denying that the rhino may have lived longer had he remained in captivity in Europe, but many conservation efforts require taking risks. Monitoring wild animals, for instance, often involves radio and satellite tracking. Attaching tracking devices to animals like lions and elephants requires sedation, which, as with humans, can be dangerous. Like conservation compromises, these imperfect solutions are the best on offer.

Not all species require active management, as Akagera's recovery shows. None has reached carrying capacity – that won't happen for quite a while, says Jes – but many species, from zebra to warthog, have been increasing in number for at least 10 years. Because Akagera has two rainy seasons, various water sources, and an abundance of plants like papyrus on which herbivores feed, growth is assured. And now that lions and black rhinos have returned, Akagera's wildlife is as diverse as it was before the country plummeted into civil unrest.

Unsurprisingly, tourism has expanded in tandem with Akagera's rewilding. It helps that, in terms of accommodation and activities, the park of today offers far more than its pre-war counterpart. In 2010, Akagera welcomed 15,000 visitors; 37,000 people visited in 2017; and in 2019, the number rose to 50,000.[14] Notably, half of the park's 2019 visitors were Rwandan nationals. Ratios like this are rare in Africa, says Naledi K. Khabo of the Africa Tourism Association. Even though parks, reserves and lodges typically offer reduced rates to domestic travellers, prices remain prohibitively high for many locals. Akagera's visitor numbers would have continued climbing in 2020 and 2021 were it not for Covid-19, says African Parks tourism manager Sarah Hall. Indeed, when I went on safari in January 2020, Ruzizi Tented Lodge, where my friend Laura and I stayed, was half-full despite it being midweek, and no rooms were free at Magashi Camp.

In Rwanda, tourism is big business: according to the Rwanda Development Board, the sector grew by around 25 per cent annually from 2013 to 2018; and in 2019, tourism revenues rose by 17 per cent, reaching $498 million. Akagera's revenues grew at an even higher rate – increasing by 1,150 per cent between 2010 and 2019 – by which time the national park generated $2.5 million a year (making the wetlands park 90 per cent self-financing). Glossy travel magazines rejoiced when Wilderness Safaris – a luxury outfitter with a conservation-minded mission – opened Magashi in 2019. *Town & Country* magazine, for instance, called it one of the best new hotels in the world in 2020.

Magashi's addition to the Wilderness Safaris portfolio is strategic. The company entered Rwanda in 2017 with the opening of Bisate Lodge in Volcanoes National Park – a property that generated serious buzz (and picked up several awards) for its unique design – and Wilderness Safaris plans to develop a Rwanda-wide circuit. Rob Daas manages the outfitter's Rwandan operations and says the company wouldn't have come to Rwanda were it not for a highly supportive and collaborative government that's pushing a 'beyond gorillas' tourism strategy.

As in many parts of Africa, Rwandans benefit from tourism in direct and indirect ways: people living near a national park receive a portion of tourism revenues generated by it, and many work in tourism and related industries. African Parks says that 80 per cent of Akagera's full-time employees are recruited from communities in close proximity to the park, and local communities receive 10 per cent of the annual revenues generated by park activities – an amount that totalled just under $160,000 in 2019. That year, staff salaries, local purchases and equipment hire amounted to $525,817. In addition, African Parks has helped launch various community enterprises – from honey cooperatives to sustainable fishing – that yield significant annual returns. More broadly, Rwanda has been hailed as an example of best practice when it comes to post-conflict resource-management and tourism.

Alphonse and Theo — the driver/guide who transported Laura and me to Akagera — have both witnessed changes in their communities based on the park's progress. According to Alphonse, communities were unhappy with early protections (like fencing) since such measures inhibited their livelihoods, but they now understand that a healthy park delivers benefits to nature *and* people. Furthermore, locals increasingly want to experience wildlife themselves. 'It's something they've been missing for a long time,' adds Alphonse.

Theo's outlook follows this trajectory. He didn't grow up with an innate interest in nature and was simply looking to improve his quality of life when he took a job working as a driver for Kigali's Hôtel des Mille Collines (the hotel featured in *Hotel Rwanda*) in 2005. Because he had a Jeep, he occasionally drove guests to Akagera, but his English was limited, so interactions were minimal. On one of his first excursions, he tossed an empty water bottle out of the window and was perplexed when his German guest reprimanded him for doing so. Theo didn't know that his action was harmful to the environment; likewise, when he began guiding, he had to learn about healthy boundaries (e.g. not getting too close to a sighting). Talking about Akagera's recovery now, he mentions the positive effects it's had on people, from job creation and tourism revenues to improved roads and new schools.

Although I'm not Big Five-obsessed, I hoped to encounter the stars of Akagera's rewilding project when I travelled to Rwanda in January 2020, but I didn't see any lions or rhinos, nor did I encounter large herds of elephants. I was disappointed to miss the animals that survived long journeys from South Africa and Europe, but delighted by the park's scenery and the smaller creatures hanging out at Ruzizi Tented Camp. I could have spent a month there watching birds. There were weavers building nests on the lakeshore, herons gripping floating papyrus and noisy, impossible-to-spot birds high in the canopy. Akagera's scenery was equally captivating, its natural beauty amplified by the seasonal presence of hundreds of thousands of yellow-and-white butterflies (the African common white, a poor name if you ask me). They flocked to rainwater pooling on rust-coloured roads and captured the attention of curious baboons.

Heavy thunderstorms – the earth-shaking kind I grew up with in Texas – precluded Laura and me from going on several game drives, and washed-out roads further restricted our activity, but on clear afternoons, we passed through western Akagera's open grasslands, pausing to photograph large herds of zebra. Later, while driving through the undulating hills of the Mubari Range that runs from north to south, we pulled over to admire a lone Maasai giraffe only to learn from our guide that the species isn't considered native to Rwanda despite living across the border in Tanzania. Akagera's population dates back to 1986 when the Kenyan government donated four long-necked ruminants.

In the context of rewilding, the giraffe's presence reminds us that baselines are only as important as we make them. Had Jes and his team pinpointed a precise date – say, 1950 – when identifying their rewilding baseline, the presence of giraffes may have posed a problem, but Jes takes a longer view. He says no one really knows whether giraffes *ever* occurred in Rwanda. The range map published by the IUCN hints at the fact that giraffes may have lived throughout sub-Saharan Africa, and the field of conservation paleobiology – which draws from paleontological and geological data – can lead to historic baselines that look very different from the ones used by the IUCN. As Julian Fennessy – co-chair of the IUCN Giraffe and Okapi Specialist Group – explains, IUCN assessments typically utilise data from the past three generations (for giraffes, that amounts to around 30 years).

Jes hopes to learn more about the historic presence of giraffes in Rwanda via electronic DNA sequencing technology. Devices that synthesise soil samples for trace DNA particles are in development and may be available for Akagera's use in the coming years. 'When it comes to finding out where species used to exist, E-DNA is the future,' explains Jes. 'But we don't want to go back 500,000 years. That's crazy.' Several thousand years (maximum) makes sense to the conservationist.

In the meantime, African Parks is focused on maintaining a biologically diverse, healthy ecosystem, and Gaël is busy documenting it via photography and writing projects.

Akagera wasn't always at the forefront of his mind, nor was Rwanda. The photographer and biologist attended university in the UK and then studied butterflies in Gabon. He saw himself staying in Central Africa for decades, but during a trip home in 2006, he noticed that Rwanda had changed dramatically since he was last resident. It showed signs of progress and was 'full of energy'. During that time, many of his peers returned from other regions full of ambition and creative ideas, but Gaël wasn't ready to move back – nor was he ready to face Akagera.

He didn't visit the park until four years later. It was October 2010: exactly 20 years after his birding club narrowly dodged the violence associated with the start of Rwanda's civil war. Gaël was pleasantly surprised to see that many animals remained in his one-time home, though he was depressed to see how some landscapes, particularly in the north of Akagera, had been cut to pieces.

When it comes to the park's wider restoration, Gaël says things are moving in the right direction, but challenges remain. 'I don't feel like it's all well and done,' he says. The park's size worries him, and he says the conversion of native to exotic vegetation taking place outside of Akagera will eventually create issues inside its boundaries. He's encouraged, however, by changes taking place in Kigali – there, people increasingly want to plant native trees instead of exotic ones. Gaël believes the tide could turn 'with the right synergies and a critical mass of interested people'.

Gaël and his father are frequent collaborators. They have co-authored four books, including *Birds in Rwanda: An Atlas and Handbook*, and their joint focus is on parks, birds and flowers. Like others in the contemporary conservation space, they hope to inspire appreciation for plant and animal species that are often overlooked. Gaël understands the importance of bringing back locally extinct species when doing so encourages ecological balance but believes species reintroductions monopolise the park's narrative. What Gaël loves most about Akagera is its 'composition: the fact that you can see so many things in one place'. Reintroductions, of course, enhance the very composition Gaël cherishes. They also reactivate trophic dynamics, in turn helping the ecosystem become healthy – and resilient – again.

CHAPTER SEVEN

The Jaguar's Journey

Iberá National Park, January 2021

Mariua took her first steps towards freedom on a rainy January night. The jaguar slipped through the open gates of her enclosure at 8.20 p.m. After walking alone for seven hours, she returned to retrieve her two small cubs and left the soft-release enclosure for a second – and final – time. Kittens on her heels, Mariua headed towards one of Iberá National Park's wetlands. Within seconds, she and her cubs passed beyond the camera's reach.

Four kilometres away, in a video surveillance centre, a group of scientists and wildlife managers cheered as they watched Mariua and her cubs confidently depart their temporary home to become Iberá's first resident jaguars in 70 years. Tears were shed as the bleary eyed conservationists mulled over the milestone's significance – the hard work that facilitated it, the hope that fuelled it and what it meant for Iberá National Park's broader rewilding.

The first of its kind in Argentina (possibly Latin America), Iberá's rewilding is the most textbook example I examine in this book. It prioritises the Three Cs – carnivores, cores and corridors – and involves highly endangered species. You'd be mistaken, however, for assuming that traditional rewilding goals make a project straightforward. In fact, if you scratch at the surface of this bold South American initiative, you'll quickly discover a labyrinth of intersecting stakeholders and protected areas.

You will, for instance, see that a conservation organisation called Fundación Rewilding Argentina manages the rewilding project, but Rewilding Argentina was created by a separate NGO, Tompkins Conservation, that also serves as its project partner. Your head starts spinning when you learn about three protected areas with comically similar names: Iberá National Park, Iberá National Reserve and Iberá Provincial Park. They sound alike but are distinct, and to make matters more confusing, the three protected areas together form what's known as Gran Iberá Park – a 709,717-hectare (1,753,749-acre) protected area that houses 4,000 species of fauna and flora, and stores an estimated 246 million tonnes of carbon. The good news is that you, dear reader, don't have to navigate the metaphorical labyrinth. That's what I'm here for. I'm your guide, and I know where to take you next.

Iberá Wetlands, 1997

Kris Tompkins peered through the window of her husband's small red-and-white plane as it rumbled over the Iberá Wetlands. Patches of flat terrain dotted with floating islands rolled by as the wetlands' 'shining waters' – the meaning of Iberá in the indigenous Guaraní language – reflected the outline of the low-flying plane. Kris was unimpressed, having grown accustomed to the picture-perfect landscapes of Patagonia, where she and her husband, Doug, lived after walking away from corporate careers in the United States. The moment she set foot on the wetlands, she thought to herself, *Where the hell are we?* 'It was hot, there were insects everywhere, and there were no landmarks to anchor us,' Kris said of her first visit to Iberá in 1997. 'The only thing I wanted to do was get back on Doug's plane and get out of there, but Doug saw something in the landscape that he never forgot.'[1]

He was taken with its raw beauty and sprawling vistas. They were worn thin by centuries of cattle ranching, but Doug could see their potential. Soon after, without telling Kris, he returned to the area and purchased the ranch that included the island, San Alonso, where he and his wife landed weeks earlier. 'That's how we got started in the conservation of the Iberá Wetlands,' says Kris. And in rewilding, as we'll soon see.

Like Greg Carr of Chapter One, Doug and Kris Tompkins were successful entrepreneurs before they entered the field of conservation. Doug co-founded two companies – The North Face and Esprit – and Kris helped create (and later ran) Patagonia, which, like The North Face, caters to outdoors enthusiasts. Doug and Kris shared a love of nature and a passion for conservation. In the 1990s, having accrued substantial wealth, the couple moved to Chile, a part of the world Doug knew well from frequent outdoor excursions. Kris, likewise, had a familiarity with South America, having spent part of her childhood in Venezuela. The duo retired from their corporate gigs, but wouldn't sit still for long.

A new chapter, this one geared towards ecological restoration, started to take shape in 1991 when Doug bought and set about restoring a farm in southern Chile. Soon, he and Kris were buying and restoring large tracts of privately owned land in the name of conservation – a model honed by The Nature Conservancy, and now deployed by numerous entities. Doug and Kris could have kept the land for themselves, hoarding its natural treasures or financially benefiting from ecotourism, but they had other plans. The American couple intended to donate the land to national governments under one condition: it had to become national parkland. Using this technique, the Tompkinses would increase the size of existing parks *and* create new ones.

Influenced by the US National Park Service, the couple believed that wilderness areas possess intrinsic *and* extrinsic value. Here it's worth sharing what US President Ulysses S. Grant said about Yellowstone National Park's 1872 creation. Yellowstone was 'set apart as a public park or pleasuring-ground for the benefit and enjoyment of the people'.[2] A lesser-known line in the storied legislation speaks to a second goal of protecting the park's 'fish and game'.[3] More controversial was the decree that all people, even long-established tribal communities, be removed from Yellowstone's boundaries and treated as trespassers should they return. 'Fortress conservation' was born.

The Tompkins philosophy contains another critical component: the belief that affluent individuals should give most of their wealth to charity. Doug, who died in a kayaking accident in 2015, walked the walk. To the dismay of one of his children (from his first marriage), his estate went to Kris and the conservation organisations they established (at the time of writing, these organisations, all of which fall under the Tompkins Conservation umbrella, have allocated $400 million to conservation). Like Doug, Kris will fight for conservation until the end. 'I'm going to do this until I drop dead,' she told a reporter in 2020.[4]

Many have praised Kris and Doug's philanthropy. The UN made Kris its patron of protected areas in 2018, and the previous year, she earned a prestigious Carnegie Medal of Philanthropy. While writing this book, I heard the Tompkins name time and again as conservationists from all over the world described their rewilding icons. Full disclosure: I'm a fan of private–public conservation models like the ones at work in Gorongosa and Iberá. Indeed, an article I wrote in 2020 entitled 'The promise of private conservation' partly inspired this book.

But the Carrs and Tompkinses of the world aren't universally admired. Some say their approach has colonial undertones; others worry it's undertaken for publicity or tax write-offs. Some critics take issue with the source of philanthropists' wealth and the environmental impacts associated with corporate enterprise. In the minds of some, charitable giving, no matter how sincere or significant, is incapable of offsetting the real or perceived damage caused during a philanthropist's corporate career. I recently came across a flurry of Instagram posts condemning the media's praise of Bill Gates's book, *How to Avoid a Climate Disaster*. One climate activist wrote, 'Stop. Uplifting. Billionaires.' On the furthest end of the spectrum is the belief that conservation as we know it is corrupted by its exclusionary history or, as one journalist put it, 'green colonialism'.[5] In an attempt to make conservation more inclusive – and fair – many modern-day conservation initiatives seek out participation from indigenous groups while launching human livelihood projects alongside species programmes.

Academics, ethicists and philosophers can (and will) debate the propriety of foreign-led conservation interventions, but, to my mind, the most relevant question is: how do people on the ground – those directly impacted by conservation and its trickle-down effects – feel about foreign conservation philanthropists like Doug and Kris? Are these outside benefactors seen as being helpful or meddlesome? Heroes or villains? One person's saviour is another's nightmare, particularly in our increasingly polarised world, but it seems to me that newspaper articles, academic journals and the like often make assumptions about the way local communities feel without stopping to ask them.

Perceptions shift like the seas, but in my experience[*] local views about foreign-born conservation leaders follow a familiar arc. Initially, communities and governments are wary of outsiders who arrive with bags of money and bold promises (see Greg Carr, Chapter One). Concerns about motivation and sincerity are particularly acute in countries that experienced colonisation. In such places, history demands that locals consider whether newcomers intend to harm or help. Let's not forget that when Spanish conquistador Francisco Pizarro reached Peru in 1532, he and his men killed thousands of Inca soldiers before taking Incan emperor Atahuallpa hostage. Pizarro promised to free Atahuallpa if a hefty ransom was paid, but he abandoned his promise after receiving boatloads of gold, instead executing the Incan leader. It was the first milestone in Spain's takeover of South America and a root cause of the Incan Empire's disintegration.

No one ran for the hills when Doug and Kris landed on San Alonso Island in 1997, but people were very sceptical of the duo. Like in Chile, where Doug got his start in the early 1990s, Argentinians struggled to wrap their heads around the Tompkins model. Locals couldn't help

---

[*] The Covid-19 pandemic prevented me from visiting a number of the projects described in these pages, which in turn limited my ability to comment on local perceptions of such projects. Of course, I interviewed as many people as possible, and I drew from past experiences reporting on (and, to a lesser extent, taking part in) conservation in various countries, from Colombia and Ecuador to Kenya and Mozambique. Still, there is no substitute for in-person conversations. My observations, then, are not as rounded as I would have liked. Take them with a grain of salt; accept or critique them, but they will hopefully spark questions in your minds, in which case I will have done part of my job.

wondering why Doug needed so much land (by 2002, Doug had acquired more than 272,000 acres of former tree plantations and cattle ranches). Rumours about the American fashion maven swirled. Did he work for the CIA? Was he acquiring land for a covert military base, or attempting to control the region's water supplies? Even level-headed locals thought Doug was *loco* when he started talking about rewilding the wetlands.

For one thing, rewilding was a foreign concept. 'Doug and Kris knew about rewilding because of their American networks, but we'd never heard of it,' explains Sofía Heinonen, executive director of Rewilding Argentina – the NGO that manages Iberá's restoration. Nor did rewilding neatly fit into Argentina's conservation agenda. 'No one was thinking about bringing species back,' adds Sofía. 'The focus has always been on habitat restoration, which remains the case in most of the national parks'.[6]

By all accounts, Doug was already thinking about species reintroductions when he purchased San Alonso in 1997, and jaguars were central to his plan, but the Tompkinses didn't immediately convey this, their boldest rewilding goal, to the people of the Corrientes province (aka Correntinos). When the (big) cat got out of the bag, Correntinos were shocked. 'We have a cattle-ranching culture, and he wants to bring back jaguars?' laughed Sofía when I asked how local people felt about Doug's plans. For reasons laid out later, Correntinos would come around on jaguars, but another of Doug's goals ruffled feathers among citizens *and* the provincial government.

He and Kris wanted the land they acquired to become a national park managed by the federal government because they considered national parks 'the most durable way to protect wildlife habitat and help people reconnect with nature'.[7] They further believed that national parks 'highlight the best a country has to show the world' and are 'proven economic drivers for local communities'.[8] Sounds like a win-win, right? Not so fast. Correntinos are fiercely independent, conservative-leaning and 'hate' the national government, says Sofía. Doug, Kris and their Rewilding Argentina counterparts would have to change their minds. Other minds needed changing, as well.

The provincial government was equally (if not more) resistant to the park's creation. Unlike local buy-in, the government's support was a necessity, not an added bonus. Because Argentina has a federal system, the central government has limited powers. National parks only come into being *if* provincial governments approve their creation. In Iberá's case, provincial support was never guaranteed. As a matter of fact, conservationists wanted to establish a national park in Iberá years before the Tompkinses turned up, but the provincial government was too proud to ratify the proposal, says Rewilding Argentina's conservation director, Sebastián Di Martino. 'Their feeling was, why does the national government need to get involved when we can manage our resources perfectly well?' Given this context, it's easy to see why locals felt like the Tompkinses didn't understand them or their culture. 'Some people thought they were crazy,' adds Sofía.[9]

For others, Doug and Kris's audacious vision was irresistible. Like a moth to a flame, biologist Ignacio Jimenez was drawn to Rewilding Iberá when he encountered a book chronicling its first decade (Doug and Kris got started in 1997, but Rewilding Argentina, the entity that implements their vision, wasn't created until 2005). The Spaniard, who became Iberá's conservation director but no longer works for Rewilding Argentina, thought to himself, 'Wow, these guys are crazy. Nobody has done this before'.[10] The more he thought about it, the more obvious it became that Doug's dream was daring but doable. Ignacio wanted in.

Austin, Texas, February 2014

Ten years ago, I moved back to Texas after almost a decade away. I missed my parents and wide-open spaces and was eager to reconnect with the family ranch after getting into wildlife conservation. My brain was overflowing with information about African wildlife, but I knew next to nothing about the plants and animals of my home state. At the ranch, I could immerse myself in a familiar-yet-foreign world while plotting ways to get the family excited about potential conservation initiatives. I tend to think five years into the future, and I had a vision – more like a fantasy – of ushering in a new era for our family, one that put conservation ahead of enterprise.

The summer before I moved down from New York City, I spent several weeks in Texas developing a potential television series, *Conservation Calling*, with hopes of spotlighting the many conservation heroes that often go unnoticed (in case you're wondering, the title likewise referred to my journey from corporate cog to wildlife activist). I decided to test the concept in Texas, so I hired a small film crew and rushed around the state interviewing dozens of conservationists. I spent a day watching aplomado falcon fledglings leave artificial nest boxes for the first time and another chatting with sea turtle rescuers. I spoke to the impassioned-but-rude owner of a sprawling animal sanctuary and visited a private ranch that breeds black rhinos. Every stop was exhilarating, but all these years later, one particular conversation sticks with me. It occurred on the family ranch, Brady, where my ragtag crew and I spent several nights.

It was late afternoon, and I was standing on the edge of Brady Creek, its emerald green waters pulsing with life, as I caught up with long-time ranch employee Jack Richardson. Like his uncle before him, Jack spent decades managing the family's cattle operations while keeping an eye on the exotic animals you met in Chapter Two. Jack and I were discussing the different ways people connect with land and wildlife. In Texas, as you might expect, many people who call themselves conservationists hunt, fish and ranch. Jack is one of them. Speaking in a thick-as-molasses Texan drawl, he told me about a theory he heard of years earlier that had resonated with him. According to it, people who hunt for sport are initially motivated by the thrill of the chase, but they evolve over time, with many gradually developing a personal philosophy that incorporates conservation – or at least stewardship.

I later reflected on the idea and how it relates to my personal evolution – not as a hunter but as someone who had a superficial appreciation for nature before I became a conservationist. On the other hand, as a child, I was more in sync with the natural world than many of my peers. I was *in* nature all the time: touching it with my own two hands and building what would become a lifelong relationship.

Not long after my conversation with Jack, my mom and uncles decided to sell the ranch. Like an activist chained to an ancient tree, I tried to stand in the way, but in the end I was powerless. In a matter of months, Brady was gone, its acreage gobbled up by a neighbouring property owner who now owns around 40,500 hectares (100,000 acres) in the area. To put that in perspective, his ranch is an eighth of the size of the state's biggest ranch and 22 per cent of the size of Iberá National Park. What a ranch of that magnitude means to him I'm not sure, but I'm quite certain he's not waiting on a 'conservation calling'.

At around the same time, a number of Iberá's ranching families grappled with the future of their land, with many deciding to sell to Doug and Kris Tompkins, in part because climate change-induced flooding made the region inhospitable to ranching. Ceding control couldn't have been easy, but Iberá's ranching families didn't lose touch with their land, not really, at least not in the way my family lost access to ours. Visiting our former ranch would require trespassing whereas families who sold their land to the Tompkinses can reconnect with their old land whenever they want – without paying a single penny.

Iberá Wetlands, 2003

By 2003, Doug had acquired 110,074 hectares (272,000 acres) of land in Iberá. Acquiring land, and lots of it, was a critical milestone in Iberá's rewilding. It paved the way for Iberá National Park, which came into being in 2018 when Tompkins Conservation donated 151,757 hectares (375,000 acres) to the Argentinian government. But land isn't the only piece of the puzzle. Rewilding science tells us that large protected areas, or cores, are only one part of the equation. After all, why set aside large swathes of land if they don't possess ecological value?

Having endured centuries of land-use change, the Iberá Doug and Kris came to know was dramatically different from the one Spaniards encountered in the sixteenth century. Once 'an untamed territory with abundant wildlife, where only hunters or explorers would venture to enter',[11] the region started to lose its wildness when European colonisation commenced in the sixteenth century. The livestock they brought with them – cattle, in particular – would forever alter the New World. But although Argentina's livestock numbers increased exponentially as Europeans laid down roots, experts say their environmental impact was limited until ranching became a driving force of landscape change in the second half of the nineteenth century.

Ranching's meteoric rise was driven by technological advancement – refrigeration was a game-changer, as were railroads – and ranching co-evolved with a growing demand for meat and livestock products in European markets. Argentina's economy flourished as an increasing number of Europeans colonised the New World, bringing with them Old World traditions and connections (in 1908, Argentina's per capita income was the seventh highest in the world[12], and when Argentina conducted its first census six years later a third of its residents were foreign-born).[13]

As ranching ramped up, regions like Iberá lost much of their wildlife. Cattle grazed vegetation away, communities killed animals for meat, and hunters cultivated feathers, fur and leather to sell to European buyers, together causing the area to experience 'one of the worst defaunation processes in northern Argentina'.[14] Numerous species vanished from the region, among them jaguars, green-winged macaws, giant anteaters and giant otters. Two of three pampas deer species vanished, too, as cougar and maned wolf numbers dramatically dwindled.

They might have joined the others in the pit of local extinction, but conservation came to the rescue. In 1983, Iberá was declared a 1.3-million-hectare (3.2-million-acre) nature reserve. With its creation, illegal hunting came under control, but farming and ranching were allowed to continue. Part of the problem lay in the fact that the national reserve included public and private land – most of which was used for cattle ranching and pine production.

If you're scratching your head, you're not alone. Why create a nature reserve if you don't intend to outlaw the activities necessitating it? And how can privately owned agricultural land also be a federally designated nature reserve? To understand this paradox, we need to look at the history of conservation in Argentina.

At first glance, the country appears to have a relatively long history of protected area conservation (the first national park – Parque Nacional Nahuel Huapi – dates back to the 1920s). However, Argentinian protected areas didn't restrict activities like logging and ranching until much later. They were 'paper parks', which in conservation parlance refers to wilderness areas with superficial protections. Legislative mandates make these empty shells of conservation look sturdy to outside observers, but closer inspection reveals a flimsy exterior and hollow interior. Paper parks are well-intentioned but poorly executed at best; at worst, they are nothing more than publicity stunts.

According to Rodolfo Burkhart of the National Parks Administration, Argentina's commitment to conservation changed in the late 1960s when Argentina began drawing inspiration from the US National Park system created a century earlier. Through a new National Parks Act, the government finally 'introduced different categories of protected areas and established national reserves, whose role was to act as a buffer zone around the parks'. Unfortunately, the legislation's impact was limited.[15] By 1990, only 22 per cent of the country's protected areas enjoyed strict protection. In 78 per cent of the country's 'protected areas' people could live, log and ranch. Paper parks were the norm, making Iberá National Reserve's paper-thin protections less shocking (albeit problematic).

New models were desperately needed. One, the rise of provincial parks managed by individual provinces instead of the federal government, pushed the needle towards meaningful conservation. Indeed, Iberá's first genuine environmental safeguards came about as a result of provincial governmental action when, in 2009, 43 per cent of the national reserve became the Iberá Provincial Park. Its 553,000 hectares (1.4 million acres) received strict protection, a move that 'enabled the slow recovery' of some of the region's wild animals, including caimans, marsh deer and rheas.[16] Things were undoubtedly moving in the right direction, but large-scale ecological restoration would require species reintroductions and other proactive measures – the kind powered by creative thinking and substantial financial capital.

Which brings us back to Doug and Kris Tompkins. Once the powerhouse couple had a number of large ranches under their belt, Doug started looking for someone to manage Iberá's restoration. Soon after, Sofía came on to his radar. The biologist from Buenos Aires was working in Iberá at the time, her focus on identifying and tackling threats to the wetlands. She knew the area – and its conservation challenges – well. In addition, Sofía had years of experience working for Argentina's National Parks Administration, which was hugely valuable: in the world of conservation, relationships can be as important as science. Doug was the visionary, explains Sofía, and his vision was crystal clear: he wanted to turn large swathes of the wetlands into a national park to ensure it was protected in perpetuity and open to the public. But first, the ecosystem would have to be restored and rewilded. The multifaceted effort required a team of experts. In short, the Tompkinses had a plan but needed people to implement it.

Doug hired Sofía in 2005, and she hit the ground running. One of the first steps involved removing unnatural species from the wetlands while putting an end to detrimental practices, such as frequent bush-burning used to artificially increase the grasses on which cattle graze. Another unnatural practice – the creation of artificial channels to irrigate rice fields – had to go since it diverted water from the wetlands. Invasive plants and trees, like *Elliotis* pine, had to be removed, and in some cases native trees were planted in anticipation of wildlife's comeback. Palm trees, for instance, were added to support the return of green-winged macaws. In the years that followed, these areas experienced what Rewilding Argentina calls a 'remarkable recovery'. Important in and of itself, habitat restoration was essential when it came to subsequent species reintroductions, particularly those involving animals like tapirs and macaws that only prosper in specific habitat types.

To get Rewilding Argentina (then called Conservation Land Trust Argentina) up and running, Sofía hired an ace, multidisciplinary team predominantly composed of Correntino natives. Next up were scientific surveys to assess the area's ecological health. Sofía and her colleagues needed to answer questions such as: which species remained in the area, and how robust were their populations? Would local habitats require a complete overhaul or straightforward restoration before Rewilding Argentina could bring back keystone species like jaguars and giant river otters?

One thing was clear: most of the species that influence the health of the wetlands – the jaguars, giant otters and macaws of our story – had little to no chance of returning on their own. Take jaguars: the closest wild population is more than 400km (249 miles) from Iberá. While jaguars can travel long distances, the landscapes between the two regions are so degraded, navigating them would be, in

Sofía's words, 'almost impossible'.[17] As we'll see later in the chapter, connectivity isn't a lost cause, but corridors aren't viable – not yet. In any case, conservationists can't afford to wait for the remote possibility that animals recolonise the wetlands in 50 or 100 years because the ecosystems *need* keystone species *now* to become healthy again. 'We don't have time to spare,' emphasises Sebastián, 'especially in light of climate change and the extinction crisis'.

Ignacio came on board in 2008 as Iberá's conservation director, and Sebastián – who manages *all* of Rewilding Argentina's conservation programmes – followed in 2015. Guided by Doug and Kris's vision, Sofía and her colleagues began brainstorming ways to bring back numerous species, jaguars included, but they were realistic about the road that lay ahead. It was long and winding and riddled with roadblocks – a marathon, not a sprint. 'When we started with the whole rewilding programme, everything was against us,' remarks Ignacio. Argentina didn't have any experience reintroducing species, nor is the practice common elsewhere in South America. 'We had to convince authorities, society and even many ecologists that it made sense to bring back extirpated species,' he adds.

Securing their support required clear, persuasive messaging. The Tompkinses and Rewilding Argentina recognised the importance of clearly communicating their conservation priorities and knew they had to carefully choose their words when doing so. 'If you tell people your work focuses on conservation, they often have a negative reaction,' says Sebastián. 'They assume you're going to restrict them from doing everything they're accustomed to, which is why we prefer the terms "nature production" and "restoration economy".'

These conceptions are business-friendly, explains the biologist, who emphasises that Rewilding Argentina and

Tompkins Conservation are working towards developing a new economy in the region, one that puts ecotourism and related businesses front and centre. According to Sebastián, this 'new economy' has already changed the way Correntinos live and work. He points to a small town near the national park where 80 per cent of residents now work in ecotourism. Some are wildlife guides; others arrange kayaking excursions. Lives are changing as new opportunities flow out of the park like a river coursing towards the sea. Most telling is the fact that many young professionals are cultivating careers close to home rather than flocking to cities, as has been the case for decades.

The symbiotic relationship between conservation and ecotourism is central to many (if not most) rewilding projects, making Iberá's 'restoration economy' a familiar tune: in beautiful, biodiverse places, wildlife tourism is a long-sought win-win. Now that (most) wildlife-lovers prefer shooting animals with cameras instead of guns, wildlife tourism can generate significant financial returns *while* protecting wildlife.

But there's a problem (there always is). Nature can't be tamed nor controlled. So-called 'acts of God' – droughts, tsunamis and pandemics – lurk around every corner. Iberá National Park has been in the works for 20 years, but it didn't formally come into being until December 2018. A year and a half later, the pandemic hit it and the rest of the world like a ton of bricks. Argentina acted swiftly, closing its borders in mid March 2020. In turn, Iberá's tourism ground to a halt as the national economy floundered (a May 2020 *New York Times* headline read, 'Argentina teeters on default, again, as pandemic guts economy').[18] Tourists from neighbouring countries were allowed to enter Argentina again from November 2020, but the border didn't begin reopening to other international travellers until 1 October 2021.

Given the newness of Iberá National Park, I can't help wondering how locals feel about wildlife ecotourism in light of the pandemic. Have they lost faith in the national park or in Rewilding Argentina for promising to deliver a restorative economy? No, Sebastián assures me. 'Local communities understand that the pandemic is out of our hands, and the provincial government has offered subsidies to the tourism sector.' Importantly, tourism did not completely dry up in Iberá. Instead, it changed. Fewer international tourists came to the park, but Argentinians sought out the wetlands more than ever before.

Spring 2006

On a warm spring day, Sofía and her new colleagues convened for the first of many brainstorming sessions. In the days and weeks that followed, a set of concrete goals coalesced. With support from Tompkins Conservation, Rewilding Argentina aimed to restore Iberá's ecosystem by bringing back animals that went extinct because of human interference. Species reintroductions were critical, as was the restoration of ecological, or trophic, interactions.

Given the complexity of multi-species rewilding, the team thought it wise to start with a single animal, and their instincts led them to a benign critter – not a predator, given their tendency to threaten livestock and people, nor a highly trafficked animal that could vanish as soon as it reappeared. They settled on the giant anteater, the largest of the four surviving species in the *Vermilingua* ('worm tongue') suborder. The animal, says Ignacio, isn't 'hated or loved by people'.

The giant worm tongue's historic range is large, extending from Guatemala to coastal Uruguay, but the species has gone extinct in at least four countries. Central

American populations are in the greatest trouble, but numerous threats – from habitat loss and wildfires to poaching for meat – have caused alarming declines in South America. Because females only give birth to one pup a year, giant anteaters are 'very sensitive to human persecution and to the loss of natural habitat'.[19] They vanished from Uruguay and went extinct in parts of Argentina and Brazil; Iberá lost its worm-tongued mammal in the mid-twentieth century. To bring the species back, Rewilding Argentina began acquiring and rehabilitating orphaned anteaters, most of which enter captivity after their mothers are killed by hunters or locals (some locals kill the animal in retaliation for attacks on dogs).

The initiative got off the ground in 2006 when a male and female were released into the Rincón del Socorro Reserve (owned by Rewilding Argentina, the private reserve abuts the land that later became Iberá National Park). When it came to public support, giant anteaters were easy-peasy, but Rewilding Argentina had a harder time working out the project's practicalities. Like the pangolin guardians of Chapter Three, caretakers at the Anteater Quarantine and Rescue Center – which Rewilding Argentina manages – had to improvise. Among other things, they had to work out the appropriate type (and amount) of food to give pups. Concurrently, Rewilding Argentina staff spent months testing and perfecting anteater-specific harnesses that would fit animals as they grew in size. A Brazilian harness was tweaked to create an Argentinian prototype.

Pups can be released once they are healthy and mature enough to navigate life in the wild – a milestone typically reached by the age of one. Before running free, they spend around 10 days in acclimation corrals within Iberá. There they become comfortable with new sights and smells while wildlife manager Rafael Abuin observes them to ensure they progress as expected. In terms of timing the release, Rafael takes his cues from the anteater, 'When we see that the animals are *tranquilo* (relaxed), we open the door and set them free.' Weekly monitoring follows, which is standard, but the methods are, shall we say, original. In the wetlands, horses take the place of cars, so when rangers like Rafael struggle to home in on an animal's location using radio telemetry, they get as tall as possible by standing in their saddles – a feat, if you ask me, as impressive as any highlighted in these pages.

Since a pair of giant anteaters reclaimed the wetlands in 2007, more than 30 have been released, and the park now contains around 150 individuals. According to Ignacio, these successful reintroductions paved the way for rewilding additional species by validating Rewilding Argentina within the conservation and scientific community. Next, the team focused on the endangered pampas deer. The once-common mammal used to roam large portions of Argentina's grasslands, but today only four isolated populations remain, each containing several thousand deer (hunting, habitat loss and disease transmitted by livestock caused the ungulate's decline).

Before Rewilding Argentina got involved, the pampas deer clinging to life in the Corrientes province lived on private lands that were fast becoming pine plantations. Translocating these animals was, in Ignacio's words, 'a rescue of sorts'. Twenty-two individuals were brought to San Alonso in 2009 to create a founder population. Over the next decade, the group rapidly grew as Rewilding Argentina launched a second nuclei in another part of the park. Iberá National Park now houses around 200 pampas deer, which amounts to a fifth of the country's total population.

Next up were collared peccaries, a herbivore that supports ecosystems by dispersing seeds. As Royal Botanic Gardens Kew explains, plants can't up and move of their own accord. They need external forces – whether strong winds, water or, in this case, animal consumption and defecation – to spread their seed. The seed-spreading peccary is an adaptable creature that, like jaguars, inhabits a large range (from the southern United States to central Argentina), but deforestation and hunting caused it to disappear from the Corrientes province during the middle of the twentieth century. Although collared peccaries occupy various habitat types, they require some shrub or tree cover to thrive, so when Rewilding Argentina started releasing the pig-like creatures into the reserve in 2015, they selected areas with ample tree cover. These days, around 150 animals live in the national park; and the birth of approximately 60 piglets highlights the programme's success.

While writing this chapter, I created a timeline to help me visualise the saga that is Iberá National Park. My North Star, the timeline draws me back to the story's central events when I get bogged down in details and lose sight of the bigger picture. Looking at it one day, I noticed for the first time that it spans 30 years on the dot. It begins with Doug's first purchase of South American land – in Chile in 1991 – and ends with the return of jaguars to Iberá in 2021. The central events, however, took place over the course of 21 years. Doug and Kris landed in Iberá in 1997, buying their first ranch the same year, and their dream of creating a national park was realised in 2018 when the Corrientes provincial government formally created the park.

It's safe to say that 1997 and 2018 were the two most significant years (not to mention the story's bookends), but

2015 stands out as well. It saw the return of peccaries, as we just learned, but 2015's importance had little to do with snouted seed dispersers. The year was significant because of two unrelated events – one joyous, the other tragic.

Joy came first, courtesy of the Jaguar Reintroduction Center, which opened following years of planning and construction. The breeding centre doubles as a training site – a place where soon-to-be-released jaguars learn to navigate life in the wild. It's where our story began, and where it ends. Its opening was momentous: a sure sign that Doug's dream of returning jaguars to the area was just over the horizon. In April 2015, Rewilding Argentina inched closer to the goalpost when the centre received its first jaguar, Tobuna, a female donated by an Argentinian zoo. Although she wasn't eligible for release – big cats that have spent significant time around people rarely are – her cubs were poised to walk free; and a potential mate, a captive male from Paraguay, had already been identified. A month later, the Rewilding Argentina team moved Tobuna from the centre's quarantine facility to a large enclosure on San Alonso, the spot where Doug landed his plane in 1997. Doug and Kris were there. They snapped photos of Tobuna, and joined community members and school children as they celebrated the milestone.

The jaguar's return was just around the corner, or so it seemed, but Doug never had the chance to observe them freely roaming the wetlands, for he died in a kayaking accident seven months later. Doug's death left a gaping hole in Iberá's rewilding programme. Plenty of progress had been made since he and Kris hit the ground running in 1997, and more plans were underway, but Rewilding Argentina would have to move forward without its visionary, its lodestar. Soon after Doug's death, Ignacio penned a eulogy for one of Rewilding Iberá's newsletters.

In it, he paid homage to Doug's vision and thanked him and Kris for helping the team develop a roadmap for the future. The roadmap, Ignacio wrote, would allow him and his colleagues to see Doug's vision through.

The next stop on the road towards rewilding involved another species reintroduction, this time of a living fossil: the South American (or lowland) tapir. The largest Neotropical herbivore has inhabited Earth since the Eocene, an epoch that ended just under 34 million years ago.[20] Having survived multiple extinction events, the four surviving tapir species face an uncertain future. The South American tapir (*Tapirus terrestris*) is considered vulnerable by the IUCN, but in Argentina it is deemed endangered. Like collared peccaries and jaguars, lowland tapirs vanished from the Corrientes province 60 or 70 years ago. Several causes are to blame, but hunting was the primary culprit. As the IUCN explains, tapirs are more vulnerable to hunting than other species as a result of their having 'long gestation and generational time'.[21] Pregnancies last 13–14 months, tapirs only have one calf at a time and calves stay with their mothers for up to 18 months.[22]

Considered a keystone species, tapirs support ecological health in several ways. Like peccaries, they disperse seeds, and a particular love of fruit has earned them the nickname 'gardeners of the forest'.[23] By digging large burrows, the massive mammals, which weigh more than jaguars, help numerous animals seeking safe havens. Finally, because they inhabit a wide variety of habitats, tapirs are 'umbrella species'. Protecting them means protecting a large suite of plants and animals, from birds to monkeys and trees.

According to Rewilding Argentina, the 2016 reintroduction (of seven tapirs) was the first of its kind in the Americas, a feat that could inspire similar efforts across the region. The team rejoiced when a calf arrived the following year. A second youngster (born to a different female) appeared in 2018, and Iberá's population continued growing until the Rewilding Argentina team discovered that some of the park's tapirs had fallen ill after contracting an invasive pathogen brought to the Americas from Africa centuries ago. Several perished; others were treated and moved into captivity. 'Because tapirs do not develop resistance or immunity against the disease and we have not found any suitable site for reintroduction in Corrientes that is free of the pathogen, we decided to remove all tapirs from the field,' writes Rewilding Argentina on its website. For now, tapir rewilding is on hold as scientists pin down the pathogen's source and develop treatment options.

When it came to reintroducing species that vanished from Iberá, jaguars always seemed like the ultimate challenge, but a charismatic parrot – the green-winged macaw – tested the rewilding team more than any other animal. Unlike jaguars, the difficulties associated with rewilding macaws had nothing to do with public attitudes towards the species. Indeed, locals are eager to hear macaws squawking around their villages since parrots attract tourists, thus generating income for community-run tourism operations. Issues stem from the fact that captive-born parrots require considerable training before building wild lives.

One of the largest macaws, the green-winged variety is what Ignacio calls a 'forest-builder' (by dispersing seeds, macaws encourage vegetation growth). The stunning parrot used to live in the forests of north-eastern Argentina, its presence noted by French naturalist Alcides d'Orbigny – who spent time in South America during the 1830s and 40s – among others. Historic records illuminate the parrot's former range while hinting at the reasons for its decline. As Rewilding Argentina explains, naturalists of the past referenced the fact that macaws were kept as pets and killed for consumption or during religious ceremonies.[24] Like big cat pelts or snake skin, beautiful feathers have long put birds at risk (side bar: the Royal Society for the Protection of Birds (RSPB) got its start in the late 1880s when several women formed groups to campaign against the use of feathers in fashion). In addition to poaching, a growing human population and rise in cattle ranching led to habitat loss and fragmentation. Together, these forces wiped out Argentina's green-winged macaws during the 1800s.

As with most of the animals returned to the wetlands, the macaws obtained by Rewilding Argentina come from breeding centres and zoos. One, Colo, was kept in a Buenos Aires flat as a pet (remarkably, he went on to become one of the most successful wild macaws). Parrots begin their rewilding journey in a quarantine facility owned by the Department of Natural Resources of Corrientes before moving to an 'acclimation aviary'. Because all of them come from captivity, they have to hone several skills before flying free. Without parents to teach them, the macaws rely on human guides like ecologist Noelia Volpe when learning how to identify native fruits, avoid predators, etc.

Although they are capable of flying around small spaces, Iberá's macaws need training to master moving between the region's isolated forest patches, which can be far apart from one another. To simulate the nuances of nature, Noelia places tree branches between the parrots' starting perch and end point – a distance of 25m (82ft). Doing so forces the macaws to make decisions about trajectories while building muscle mass. She also helps them learn to identify predators by playing audio recordings of macaw warning calls while a trained cat attacks an embalmed macaw in the parrots' vicinity. Once captive green-winged macaws pass these tests, Noelia takes them out into the open. Using whistles and food stations located high in the trees, she helps her trainees safely fly longer distances.

Finally ready for life in the wild, a cohort of 13 green-winged macaws flew out of their temporary homes and into the open wetlands on 29 October 2015 – at least a hundred years since the bright-winged bird vanished from the area (and likely the country). All reintroduced parrots wear small radio transmitters, allowing the Rewilding Argentina team to monitor their movements. At times, this omnipresence comes at a cost. Deaths have occurred in most of the species returned to Iberá, however green-winged macaws have fared the worst. Within a year and a half of the autumn 2015 release, five of 13 parrots had perished and none had successfully reared a chick, making the population's reproductive rate zero and mortality rate 46 per cent.[25] All but one of the parrots died as a result of predation, but in one instance locals fed a young macaw seeds the bird couldn't digest (a public awareness campaign against feeding wild animals has since been launched and Noelia continues to use various techniques to train captive parrots how to avert predators).

Reintroductions are inherently risky: not all succeed, and those that do often come at a cost.[26] Some conservationists would look at these rates and the time and resources required to prep parrots for the wild and call the programme a failure. Rewilding Argentina doesn't see it that way. Project managers underscore that once a pair unites, it takes a long time for them to establish a territory. Several breeding seasons typically pass, they add, before a couple is capable of successfully rearing chicks. Recent developments bear this out: one breeding pair laid fertilised eggs in 2019, but the eggs broke several days later, probably as a result of the inexperienced adults accidentally crushing them.

*

Then, in October 2020, the pair welcomed three hatchlings – the region's first wild-born macaws in more than a century. Given the parents' lack of experience, the team closely monitored the situation to ensure chicks were adequately fed. If they were underweight during health checks, the team provided supplemental food before returning chicks to the nest. We know from other chapters that intensive management is often required in novel rewilding projects, but we also know that it can't last forever without compromising the 'wild' in rewilding. Today, two of the hatchlings fly free alongside 17 adults, all of which freely move through the wider ecosystem. Their vibrant hues have been spotted throughout the national park and in villages and on cattle ranches.

In the midst of restoring healthy habitats and reintroducing animals that eat insects, fruits and plants, Sebastián and co steadily worked towards their goal of re-establishing a complete food chain dominated by carnivores. All of the animals reintroduced to Iberá contribute to the region's ecological health, but apex predators are the most important. The principles of rewilding tell us so: top predators 'influence their associated ecosystems through top-down forcing and trophic cascades, which in turn lead to myriad effects on other species and ecosystem processes'.[27] One study evaluated wildlife density and diversity on several newly formed islands created by the construction of a hydroelectric dam in Venezuela. Scientists found that within a very short period of time (seven to eight years), islands devoid of jaguars and pumas lost up to 75 per cent of their vertebrate species.[28] In the absence of large predators, rodents, birds and other animals that consume invertebrates and seeds exploded in number. Things were, to put it simply, out of whack.

Conversely, apex predator reintroductions reactivate trophic interactions by consuming animals lower on the food chain, which in turn causes a cascade of interactions between organisms on all parts of the chain. In 1995, wolves were reintroduced to Yellowstone National Park. Decades later, scientists were able to demonstrate that the predator's return sparked a series of changes that recalibrated Yellowstone's ecology. Iberá's rewilders hope to document a parallel transformation in the wetlands, but first they need to establish a robust population of jaguars.

As apex predators, jaguars occupy the top of the food chain throughout their American range, which starts in the middle of Argentina and ends in the south-western United States. The American population is negligible, however: individual sightings are exceptionally rare and only involve males venturing north from Mexico in search of mates. Unfortunately, they're looking for love in all the wrong places since the United States hasn't contained a breeding population of jaguars in 50-odd years (numerous scientists are pushing for their reintroduction in the American Southwest,[29] but resistance to apex predator reintroductions elsewhere in the United States suggests the dream may not materialise). In better news, jaguars have 'only' been declared extinct in two countries – El Salvador and Uruguay – but the New World's biggest cat has lost 40 per cent of its historic habitat. That percentage jumps to 95 in Argentina. Mariua and her cubs aside, Argentina's jaguars only survive in three distinct patches in the north of the country, which is highly problematic given that jaguars' long-term survival requires connectivity.

The usual culprits – hunting, poaching and habitat loss – are to blame, though the trade in big cat body parts mentioned in Chapter Six has spiked in recent years, amplifying concerns among big cat conservationists. On the plus side, hunting for pelts has decreased since the 1970s.

As Doug Tompkins knew from the start, jaguars are indispensable to Iberá's rewilding. Acknowledging this fact is one thing; acting on it is another. No one had ever tried reintroducing jaguars before and, as mentioned previously, Argentina had no experience with rewilding, making the initiative far more challenging than it would have been in countries like South Africa where large mammal translocations have occurred for decades. This lack of experience impacted the work itself – Rewilding Argentina needed to gain experience before taking action – and influenced the organisation's ability to secure support from local communities and government authorities. As we see time and again, human coexistence with large carnivores is 'essential for rewilding'.[30]

To gather experience, Argentina's fledgling rewilders turned to gurus elsewhere. 'Whenever we start a new project, we seek out people who can help us,' explains Sebastián, but 'we're interested in practitioners, not academics. Often times, individuals who are too academic or science-y don't have useful advice, and they want to see a ton of data before making decisions.' According to Sebastián, Argentinian practitioners are 'very rare', so he cast a wide, and global, net. Brazil was an obvious starting point. Brazil and Argentina share many of the same species, and Brazil has what Sebastián calls a tradition of 'active wildlife management', so he and his colleagues invited numerous Brazilian conservationists to visit Iberá.

However, Sebastián was determined to meet the 'kings' of big cat rewilding, and doing so meant travelling to Africa. I was not at all surprised to learn that he travelled to Phinda (Chapter Four) and consulted African Parks (Chapter Six). Connections like these pop up throughout these pages – a reminder that, in spite of its popularity, large-scale rewilding is a relatively niche endeavour. Links between projects tell us something else: they demonstrate that a vibrant global exchange – of rewilding information and inspiration – is taking place as we speak.

In South Africa, Sebastián met with wildlife translocation master Grant Tracy, who has been capturing and transporting big game for 30 years through Tracy & Du Plessis Game Capture. Grant taught Rewilding Argentina's staff ethical capture techniques, and introduced them to top-of-the-line translocation trailers and pens that influenced the ones Rewilding Argentina later built.

Knowledge is power, but every region (and species) is unique, so Rewilding Argentina couldn't just copy and paste foreign models when bringing jaguars back. Adjustments would be required, new tactics developed. Moreover, Sofía and her colleagues still needed to secure support from local communities and the provincial government. Here, communication is everything.

Around a million people live in Argentina's north-easterly Corrientes province. The 8.9-million-hectare (22-million-acre) province sits in the country's Mesopotamia region and shares borders with Brazil, Paraguay and Uruguay. Centuries of cattle ranching led to 80 per cent of the region's surface being converted to ranch land, but although ranching is central to the region's economy and ethos, Correntinos aren't cowboys through and through. Provincial culture is heavily influenced by the fact that many locals possess Guaraní ancestry and continue to identify with their indigenous heritage, and jaguars are Guaraní icons. The word 'jaguar' stems from the Tupi-Guaraní word *yaguára* or *yaguareté*,[31] and jaguars frequently appear in Guaraní folk music (*chamamé*) and mythology.[32]

Ignacio, Sebastián and Sofía could have extrapolated from these traditions, assuming that locals would be in favour of living side by side with jaguars, but their scientific backgrounds – and an abundance of caution – inspired a data-driven approach. In 2010, Ignacio surveyed just under 450 local men and women to evaluate their attitudes towards jaguar reintroductions, finding that 95 per cent of respondents supported the big cat's return.[33] Notably, a quarter of respondents listed their heritage as the main reason for supporting the initiative (44 per cent cited tourism potential).[34] According to Ignacio, the jaguar's strong and independent nature is another key factor, 'Correntinos may consider the jaguar as a powerful metaphor of their own character because it is Guaraní, and therefore authentic, and it is also aggressive and "guerrero" (warrior-like), as are the Correntinos themselves.'[35]

Iberá National Park, April 2021

Four months have passed since Mariua and her two small cubs walked away from their enclosure on San Alonso in search of a wilder life. Mariua made her first kill – a meaty capybara – within a matter of days. She has kept her cubs safe and helped them grow into vibrant eight-month-olds. Assuming they stay healthy and avert human-made risks, they will eventually peel away from their mother and build solitary lives of their own, as nature intended.

The family's immediate success suggests that Rewilding Argentina's approach to jaguar rewilding is working, but the family of three can't spawn a sustaining wild population on their own. Unfortunately, given the newness of jaguar reintroductions – not to mention the fact that captive-born big cats are rarely releasable – Rewilding Argentina has limited options when sourcing rewilding candidates.

Tobuna, the breeding centre's first resident jaguar, never birthed cubs, but she played a critical role before retiring several years ago. Her contribution to the project occurred in El Impenetrable, a national park that is located in Argentinia's Chaco province. Created with support from Tompkins Conservation in 2014, the national park is the largest in northern Argentina. In September 2019, Sebastián translocated Tobuna to an enclosure in El Impenetrable with hopes of attracting a wild male that had recently appeared out of thin air. A ranger was astonished when he came across jaguar tracks, and soon remote cameras confirmed the jaguar's presence.

They named him Qaramta ('the one who cannot be destroyed') in a nod to his bold nature and good fortune (Qaramta, who wandered into Argentina from Paraguay, frequents unprotected areas where logging and hunting threaten his survival). Jaguars were presumed extinct in the park, so his presence warranted celebration, but scientists needed to capitalise on his genes by encouraging him to mate with Tobuna. Like Justin Chuven of Chapter Two, Sebastián played matchmaker as the team tested the 'highly creative solution'[36] of using Tobuna as a sort of mating bait. Even if the two cats didn't copulate, her presence might anchor Qaramta to the park, preventing him from slinking back to Paraguay.

It worked. Qaramta spent more and more time in the vicinity of Tobuna's enclosure, which gave Rewilding Argentina the chance to radio-collar him. Meanwhile, Sebastián's team furiously built a special breeding pen that would accommodate the wild male and a captive female. In the end, Tobuna and Qaramta weren't a match – the 17-year-old female never became pregnant despite several mating attempts – but Rewilding Argentina had a back-up plan. They brought Tobuna's daughter, Tania, to the facility. She and Qaramta mated in late 2020, not long before Mariua and her cubs walked free, and three months later, Tania gave birth to two healthy cubs. They are believed to be the first ever jaguars produced by a wild/captive pairing. Another first in a story composed of record-breaking moments.

For the next few years, Tania's cubs will learn invaluable life skills from their mother while growing up in a large enclosure located inside El Impenetrable. As in Iberá, human interaction will be kept to a minimum to ensure the cubs can return to the wild when ready. Rewilding Argentina plans to send them bounding into El Impenetrable, thus establishing another jaguar cluster within the country.

In the meantime, Iberá's jaguar population continues to grow. During April 2021, three jaguars joined Mariua and her cubs in the vast wetland park. The newly released female, Juruna, is Mariua's sister, making the park's four cubs cousins. It's tempting to imagine a Disney-like family reunion replete with chuffing and cheek rubs, but jaguars' solitary nature means a friendly meet-up is unlikely (though possible; related big cats have been documented greeting one another with apparent recognition). Cubs stay with their mothers for several years, at which point they strike out on their own, and adult males only spend time with other jaguars when mating. Besides, Iberá's large size means the families may not cross paths. However, their impact will be collective.

Sebastián likens them to the wolves that rebalanced the Yellowstone ecosystem. Jaguars eat various prey species, however Sebastián anticipates them mainly dining on capybara – a large, grazing rodent that in the absence of predators has overtaken parts of Iberá. More capybaras mean less vegetation, which in turn impacts a variety of species (and interactions). Given how recently jaguars returned, scientists have not yet documented the apex predator exerting 'top-down' control on the food chain, but they are actively studying how reintroduced animals interact with one another and their new environment. Already they have observed interesting dynamics, from peccaries eating ticks off capybaras to vegetation changes caused by animal foraging. Sebastián is delighted by the recovery of these inter-species relationships. He adds, 'Because we have a good baseline, we can measure changes over time.'

You may be surprised to learn that jaguars are not the only top carnivores that vanished from Iberá. The region (and country) also lost the endangered giant otter – a river-dwelling apex predator pushed out of Argentina during the mid-twentieth century as a result of hunting, development and dams. Bringing otters back is part of Rewilding Argentina's plan to make Iberá 'whole and functional from an ecological perspective', says Sebastián. As seen in the Brazilian Pantanal, the fish-eating mammals are tourist attractions since they're active during the day, charismatic and live in family groups, but their position as the region's top aquatic predator is their most valuable asset, underscores Rewilding Argentina. Several otters are on-site and waiting in the wings, but none have been reintroduced yet. Because giant otters live and hunt in family groups, Sebastián is waiting for the park's existing breeding pair to have pups so he can send a unified family into the wetlands.

However, something remarkable happened during May 2021 that could change the fate of Iberá's otters. Sebastián was kayaking in El Impenetrable when he noticed an animal swimming across the Bermejo River. All of a sudden, its head and chest popped up, allowing Sebastián to see that he was staring at a giant river otter – the first observed in the Argentinian wild in four decades. The

biologist suspects the lone individual made his way to El
Impenetrable from Paraguay (the closest known population
is 1,000km (621 miles) away), but he acknowledges that
Argentina *could* contain a small population that has gone
undetected for years. In either case, the otter's presence
demonstrates that nature is resilient.

Twenty-five years have passed since Doug and Kris
Tompkins landed in the Iberá Wetlands for the first time.
The visit could have been their last – a blip on the radar for
the powerhouse couple, and the region – but Doug saw
something in the wetlands that day, something that lured
him back time and again. Like local conservationists before
him, he looked at rundown ranches and saw a stunning
national park. He observed cattle but imagined jaguars.

Together, he and Kris launched one of the most
ambitious rewilding initiatives in the world. By the end of
2020, Tompkins Conservation had ensured the permanent
protection of 5.7 million hectares (14.2 million acres) in
South America with the help of implementing partners
like Rewilding Argentina. The organisation started by
Doug and Kris has helped build or expand 13 South
American national parks, Iberá included – a colossal effort
that saw the couple pour nearly $400 million into
conservation. If you look at the project map on the
Tompkins Conservation website, you see a string of
protected areas starting with marine parks south of Cape
Horn going all the way up to El Impenetrable and Iberá in
the north of Argentina.

Another map produced by Protected Planet underscores
the importance of Tompkins Conservation protected
areas to the countries surrounding northern Argentina
(from west to east, Bolivia, Paraguay, Brazil and Uruguay).

The IUCN- and UN-managed website contains an interactive map showing every protected area in Latin America and the Caribbean. Importantly, the map distinguishes between paper parks and areas possessing effective management. When you zoom in to northern Argentina, you see that Iberá and El Impenetrable are two islands of green (the colour demarcating effective protected areas) in a sea of beige (the hue representing unprotected terrain). They are safe harbours – and stepping stones – that will become more significant as climate change and human population growth cause suitable habitat to further shrink.

Conservationists are desperate to increase the size of global protected areas. Bold pledges like '30 by 30' – a country's goal of protecting 30 per cent of its terrestrial and marine areas by 2030 – may well fail. A similar target set in 2010 as part of the United Nations Decade on Biodiversity that aimed to protect at least 17 per cent of the planet's terrestrial areas by 2020 fell short by 2 per cent.[37] Failures like these aren't all bad: they incentivise people and policy-makers to aim higher and go further.

Even with expanded protected areas, animals like jaguars require connectivity to move between cores. Whether migrating, searching for sustenance or seeking love like the lonely jaguars crossing the US–Mexico border, animals with large home ranges need wildlife corridors to get from point A to point B. Conservationists have long recognised the need for jaguar corridors, and some have attempted to create them. In 1990, several conservation organisations banded together with hopes of creating the *Paseo Pantera* ('Path of the Panther'), which would have run from southern Panama to south-eastern Mexico. For reasons too complex to mention here, the project stalled without having made much progress.

Tompkins Conservation and its partners know connectivity is critical and have begun identifying areas where they can support corridors, but the last of the Three Cs is the most long-term goal of all. 'We may not see functioning corridors in our lifetime,' says Carolyn McCarthy, who manages the organisation's global communications, 'but we're planting the seeds now.'

CHAPTER EIGHT

Replacement Rewilding

Hence, both in space and time, we seem to be brought somewhere near to that great fact — that mystery of mysteries — the first appearance of new beings on this earth.

Charles Darwin, *The Voyage of the Beagle*

Finches are on my mind the second I exit the plane and step on to balmy Baltra Island in the Galápagos. My fellow passengers hurry towards immigration, but I linger outside, scanning shrubs for any sign of the birds that helped Darwin develop his theories about evolution.

They aren't beautiful, as Darwin himself noted when visiting the archipelago in 1835, nor should they be. Perfectly adapted to the volcanic island chain, all 18 species of Galápagos, or Darwin's, finch are drab. 'Most of the peculiar Galapageian species' are 'more dusky coloured' than their mainland counterparts, Darwin lamented in his journal. Evolutionary explanation notwithstanding, Darwin frequently ridiculed the lacklustre animals he encountered during five weeks of fieldwork. He was especially critical of the world's only marine iguana, calling it 'hideous-looking', and Galápagos landscapes rarely impressed the English scientist. Describing his initial observations (on San Cristóbal Island), Darwin wrote, 'Nothing could be less inviting than the first appearance.'

Several days later, during a 'fatiguing' walk across the island's volcanic craters, Darwin finally experienced something resembling delight when he encountered a 'strange Cyclopean scene' starring two giant tortoises.

'These huge reptiles, surrounded by the black lava, the leafless shrubs and large cacti, seemed to my fancy like some antediluvian animals,' wrote Darwin in his notes. He frequently climbed on their shells, tapping to compel movement, and might have spent all day riding around like a child on a pony had he not 'found it very difficult' to keep his balance. Unfortunately, Darwin spent more time playing with (and eating) tortoises than studying their adaptations – despite the fact that then-governor Nicholas Lawson, a non-scientist, told Darwin that tortoise size and shell shape varied by island. It was only later, when Darwin reflected on his observations and Lawson's words, that he experienced the aha moment for which he is known.

Like his finches and the islands' four endemic mockingbirds, giant tortoises helped Darwin see the splintering of species that has occurred during the Galápagos' relatively short history (the volcanic islands burst through the water's surface between 5 and 10 million years ago). Before noting minor variations among features like finch beak size, Darwin 'never dreamed that islands, about 50 or 60 miles apart ... formed of precisely the same rocks, placed under a quite similar climate ... would have been differently tenanted'. As it turns out, oceanic islands like the Galápagos and Hawaii are known for containing a high number of similar-yet-distinct species, thanks to a process called 'adaptive radiation'. In his most famous work, *On the Origin of Species*, Darwin finally articulated the reason behind these minor modifications, writing, 'Species occasionally arriving after long intervals in a new and isolated district, and having to compete with new associates, will be eminently liable to modification, and will often produce groups of modified descendants.'[1] He was, of course, describing evolution. You may be surprised to learn that Darwin didn't develop the revolutionary theory all by himself, nor did it spring into his mind as he tromped around volcanic island craters. However, his experiences on the remote archipelago, and on the HMS *Beagle*, were 'the transformative period of Darwin's life'.[2]

Darwin may not have recognised evolution straight away, but he immediately noticed that in the Galápagos, reptiles reign supreme. He wasn't the first to make the observation: the Dominican friar who discovered the islands in 1535 named them after the old Spanish word for saddle in a nod to the shape of saddle-backed Galápagos tortoises (five of the archipelago's 15 giant tortoise species have saddle-shaped shells while ten are 'domed'). Darwin wrote, 'there is no other quarter of the world where this Order replaces the herbivorous mammalia in so extraordinary a manner'.[3] Like the kākāpō of New Zealand, the giant tortoises and iguanas that alternatingly amused and disgusted Darwin fill ecological niches ordinarily occupied by mammals. In case you're wondering, the Galápagos Islands contain six native terrestrial mammals, including two types of rat, versus New Zealand's three native land mammals, all of which are bats.

You will recall that New Zealand's dearth of mammals stems from the fact that the country split off from supercontinent Gondwana before mammals evolved, but the Galápagos Islands have never been attached to continents, so their general lack of land mammals has a different explanation – one relating to transport. Species native to oceanic islands didn't appear out of thin air – they had to travel from nearby continents to their new island homes. How did they do so? They flew, floated or hitched rides on rafts. Long-distance journeys of this kind are infrequent but still occur today. In the mid 1990s, a batch of 15 green iguanas turned up on a new Caribbean island after travelling 322km (200 miles) on a raft of vegetation.

The Galápagos and other oceanic islands contain what author Jerry Coyne calls unbalanced biotas (not to be mistaken for unbalanced, unhealthy ecosystems). Coyne is referring to the fact that on oceanic islands certain species types are missing while others proliferate. As he explains, 'Although oceanic islands lack many basic kinds of animals, the types that are found there are often present in profusion, comprising many similar species.' Galápagos tortoises perfectly illustrate his point. The 15 species – several extinct, most extant – share a common ancestor. Scientists believe that a single individual (a pregnant female) or a mating pair made their way from South America two or three million years ago and founded the resulting population. Tortoises first colonised the eastern islands of Española and San Cristóbal before dispersing to 10 of the archipelago's largest islands. Although scientists don't have a name for the species that colonised the Galápagos, they are confident about the fact that the Galápagos tortoises' closest living relative is the medium-sized Chaco tortoise of South America.

In the Galápagos, giant tortoises are keystone species that act as ecosystem engineers. Not unlike elephants, the large reptiles move seeds between areas by eating in one place and defecating in another (their poo, helpfully, contains a large number of intact seeds). Digestion takes time, and tortoises move around as plants pass through their digestive tract. A 2012 study found that, on average, giant tortoises travel 394m (1,292ft) during 12-day digestive periods, but one poor soul ventured 4,355m (14,288ft) during a 'retention time' of 28 days.[4] In addition to seed dispersal, giant tortoises engineer ecosystems by trampling vegetation. Both services alter plant communities, a process that in turn benefits other animals. More recently, scientists have begun investigating ecological impacts associated with the giant tortoise's penchant for wallowing in ponds. They suspect that tortoises, like hippos, fertilise ponds via defecation, but have to conduct more research before making any conclusions.

Giant tortoises have engineered ecosystems for millions of years – and not just in the Galápagos. Now relegated to the Ecuadorian archipelago and several islands in the Indian Ocean, giant tortoises used to live on every continent except Antarctica, so although they seem uncommon to us, they were once widespread. Scientists estimate that at least 36 species of large and giant tortoise vanished during the late Pleistocene – a period pinned to 11,700-129,000 years ago. As we see time and again, the qualities that set giant tortoises apart led to their downfall. Hard shells protect them against most predators – but not humans – and their ability to survive for months without food or water serves them well in harsh, arid zones. It also explains how giant tortoises survived while drifting from Ecuador's mainland to the Galápagos (several weeks would have passed as they travelled the 600 miles).

Unfortunately, this superpower made giant tortoises attractive to seamen seeking long-lasting food. The Galápagos weren't *settled* until 1807, but humans first *arrived* in 1535. Considerable damage began during the seventeenth century, when pirates used the Galápagos while waging attacks on Spanish convoys, and whalers began turning up towards the end of the century. Whalers decimated the region's sperm whales as fur traders killed tens of thousands of Galápagos fur seals. Buccaneers and whalers greedily loaded tortoises on to their boats as they prepared for months at sea. Sometimes they took too many, but lightening their load was easy: they simply tossed live tortoises overboard. As we see later, the cruel practice had unintended consequences that now benefit giant tortoise conservation.

Mariners captured tortoises for meat, but the hard-shelled reptiles were also exploited for their oil, which powered street lamps in cities like Quito. In his journal, Darwin touched on large-scale poaching and the reasons behind it. The naturalist noted that giant tortoise numbers were 'greatly reduced' by the time he visited the Galápagos in 1835, but in the same sentence he observed that locals could 'count on two days' hunting giving them food for the rest of the week'. Darwin's lack of alarm over human consumption suggests that giant tortoise populations may have *seemed* robust, or at least stable, during the mid-nineteenth century. In fact, by then, much of the damage was done.

Within two decades of Darwin's visit, the Floreana giant tortoise went extinct, and the Santa Fé variety vanished at around the same time. Pinta tortoises came next, deemed extinct in 1906, and other species were in steep decline, shielded only by the fact that their islands had not yet been infiltrated by people. Ultimately, upwards of 200,000 giant tortoises were killed or removed from the Galápagos. Floreana was the first island colonised by people (in 1807) and the place where Ecuador officially laid claim to the Galápagos in 1832. The island's early residents were a colourful bunch. There was a drunk Irishman who 'struck every one with horror', criminals and political prisoners, and a self-proclaimed baroness who mysteriously disappeared, and was likely murdered, several years after arriving.

The sixth largest island in the Ecuadorian archipelago, Floreana is the smallest of four islands inhabited by people. It spans 17,300 hectares (42,749 acres), which is roughly a ninth of London's size. It is unlikely that more than several hundred people ever lived on Floreana at any one time, but the men and women who set foot on the island, whether for brief stays or the long haul, left an indelible mark. One man – a sailor aboard a Nantucket whaling ship – destroyed huge swathes of Floreana in 1820 when he set fire to a portion of underbrush in a poorly conceived prank. It quickly grew out of hand and was still smouldering the following day as his vessel sailed away. The damage was immense – unquantifiable – and may have caused local extinctions.

In the decades that followed, invasive plants and animals infiltrated Floreana as Ecuadorian and European colonists put down roots. Cattle, donkeys, goats and pigs arrived, as did cats and dogs. Invasions occurred across the archipelago: over time, people introduced some 30 vertebrate and 536 invertebrate species to the Galápagos, as well as 870 foreign plants. In tandem, agricultural activities altered the island. A 2010 study revealed that agriculture transformed 48 per cent of Floreana's arable humid zone. Other islands have fared even worse: on San Cristóbal, 90–100 per cent of the humid highlands have been changed; and on Santa Cruz, that figure is between 76 and 88 per cent.[5]

We learned in Chapter Three how invasive mustelids and rats threaten New Zealand's endangered, ground-dwelling birds, but what about the alien animals of the Galápagos? How have they harmed the giant tortoises (and other uniquely adapted species) that managed to withstand centuries of direct exploitation by humans? The answer depends on the invader. Goats graze on the vegetation giant tortoises consume, making the animals 'resource competitors'. Unlike long-lived, slow-paced tortoises, goats breed on a fast timescale. As a result, populations grow exponentially (according to the Charles Darwin

Foundation, populations can increase by 50 per cent each year). As goats edged out tortoises, plant communities changed, with woody vegetation dominating vast areas. Livestock inadvertently trample tortoise nests while ants, rats and pigs consume eggs. Feral dogs, meanwhile, are known to attack adult tortoises. And infrastructure – roads, fences and the like – impede tortoise migration.

In 1934, the Ecuadorian government enacted legislation protecting a large section of the Galápagos Islands and their native fauna and flora. It was a revolutionary first step towards conserving the archipelago. The next major milestone occurred in 1959 when the government set aside 97 per cent of the archipelago's landmass as the country's first national park. The year was symbolic: 100 years earlier, Darwin's *On the Origin of Species* was published for the first time. As author Henry Nicholls writes in *The Galápagos*, the anniversary 'provided the necessary impetus to do something with the islands'.[6] 'The link with Charles Darwin,' he adds, 'gave the Galápagos a certain symbolic appeal in the minds of the pioneering conservationists of the 1950s.'

The same year the national park came into being, scientists established the Charles Darwin Research Station on Santa Cruz Island. Although scientific expeditions began decades earlier, the research station was the archipelago's first permanent science hub. Given the remoteness of the Galápagos, and the fact that scientists from all over the world have contributed to understanding and protecting the archipelago, a permanent base was a game-changer. The research station's first major contribution came in 1965 with the launch of the inaugural giant tortoise breeding centre on Santa Cruz Island.

Giant tortoises were an urgent priority when conservation began gearing up in the 1950s and 1960s, since three of the 14 originally named species had gone extinct. Others were in varying states of decline. The centre's premier programme focused on the critically endangered Pinzón Island giant tortoise. Only 200 adults remained, and no young tortoises could be found. Black rats inadvertently introduced to Pinzón towards the end of the nineteenth century consumed hatchlings, thus halting population growth. Giant tortoises are among the planet's longest-lived creatures, with some having reached 175 years of age, so scientists had time to rebuild the population using captive breeding, but whether they could clear the island of its troublesome invader was a separate question.

The situation on Española Island – one of the oldest in the archipelago and a frequent stop for sailors – was even more dire. Once estimated at 5,000, the island's tortoise population had tumbled to 15 in 1960. So, starting in 1963, conservationists moved the survivors to the Santa Cruz breeding facility. Fortunately, the group consisted of more females (12) than males (3), which gave the population a higher chance of rebounding. Captive hatchlings appeared in 1970, and the first repatriation of captive-born Española tortoises took place in 1975. The first *wild-born* hatchlings were documented in 1990. Scientists anticipated this gap since giant tortoises don't start reproducing until they are at least 20 years old. Although the programme found success early on, only two of the three males transferred from Española were capable of breeding, so the centre acquired a third male, Diego, from the San Diego Zoo in 1977.

Diego's origins are unclear, but he was likely taken from the wild as a scientific specimen during his youth. One thing *is* clear, however: Diego is a rock star breeder with a 'species-saving sex drive'.[7] I had the chance to see the randy reptile while visiting the Santa Cruz Breeding Center in 2019. Although I caught numerous giant tortoises in the act during my tour of the centre, Diego was taking a hard-earned break from breeding when I spotted him. A year later, he officially retired after siring nearly 2,000 young tortoises during his 40-year tenure as a full-time Romeo. Together with captive breeding experts and dedicated keepers, he and his fellow breeders brought Española's wild population up to 3,000 from 15. Amazingly, the island now contains 40 per cent of its historic population.

Diego and his libido became famous, but another male tortoise was even better known. George, or Lonesome George, as many of us knew him, became an inadvertent mascot – first for Galápagos tortoises and later for extinction.

On 1 December 1971, Hungarian biologist József Vágvölgyi walked around Pinta Island in search of snails, his speciality, when he suddenly came upon a giant tortoise. Unbeknownst to Vágvölgyi, Pinta tortoises were presumed extinct (scientists removed the last-known individual from the island in 1906). Without realising the significance of his sighting, Vágvölgyi carried on searching for snails, but first he snapped a photo of the tortoise that came to be known as George. News of Vágvölgyi's remarkable sighting soon made its way to Peter Kramer, then-director of the Charles Darwin Research Station, and not long after, a team of scientists and park wardens journeyed to Pinta to look for George. It took time, but the lone (and lonesome) tortoise was eventually located on the island's western slope. Like the pirates and whalers who stole George's kin, park wardens had to manually carry George to a boat bound for Santa Cruz.

Determining the age of an adult tortoise is difficult, but scientists estimated that George was likely around 60. This was fortuitous. It meant that George had the potential to become the next Diego *if* a mate could be found. Like Diego, hundreds of tortoises were taken from the archipelago during scientific and zoo-collecting expeditions launched during the first half of the twentieth century. The largest was carried out by the California Academy of Sciences in 1905–6. It yielded 78,000 individual biological specimens, one of which comes into play later in the chapter. In 1928, the New York Zoological Society, which appeared in Chapter Two and later became the Wildlife Conservation Society, sent a crew to the Galápagos to retrieve tortoises for conservation-oriented captive breeding in the United States. A mighty 182 tortoises were collected and distributed to facilities like the Bronx and San Diego Zoos (15, I learned, went to the San Antonio Zoo). It was possible, then, that George was not the last of his kind, but hope was in short supply. The press dubbed George 'lonesome' and the name stuck.

As scientists scoured American zoos for tortoises with Pinta ancestry, keepers at the Santa Cruz Breeding Center tried pairing Lonesome George with female Española tortoises (an Española/Pinta hybrid wasn't ideal, but it was better than nothing). Unfortunately, George showed little interest in mating. Who could blame him? As a Swiss biologist who spent years working with George explained, 'He probably never saw a female and male of his own species reproducing.' The team brought breeding males into the pen George shared with several females to show him how it was done, but George was uninspired. Attempts to collect semen samples were equally futile. George, it seemed, was destined to remain lonely.

Meanwhile, conservationists continued breeding Española tortoises and launching ecological restoration projects. As in New Zealand, Galápagos island restoration began with the removal of invasive species. In the Galápagos, goats are enemy number one. They reached the Galápagos during the 1800s and eventually colonised 13 islands. Because goats will eat pretty much anything, they are able to rapidly decimate vegetation. As the Galápagos Conservancy explains, feral goats 'are particularly devastating to island ecosystems, causing ecosystem degradation primarily through overgrazing, destruction of forests, and causing erosion'.[8]

Goat eradications began in the Galápagos soon after the archipelago became a national park in 1959 and ramped up in the 1970s. During that decade alone, more than 40,000 goats were shot, and by 1997 the national park had managed to clear five small islands, however on medium and large islands, small numbers of goats continually evaded culls (which is unsurprising when you consider that Lonesome George went unnoticed for decades). It soon became clear that removing alien ruminants from the Galápagos required creative tactics.

Inventive tactics were first trialled on Pinta Island in 1999 as part of a wider eradication effort, Project Isabela, that targeted the northern part of Isabela Island, all of Santiago, and all of Pinta. By this time, park rangers had spent decades attempting to rid Pinta's 5,940 hectares (14,678 acres) of goats. In fact, they were scheduled to visit Pinta for a cull when they learned about Lonesome George, giving the trip dual purposes. To locate elusive goats, rangers deployed what they called 'Judas goats'. The national park's secret weapons weren't hardwired for betrayal nor bioengineered to attack. Betrayal, for them, was achieved without malice or knowledge. Sterilised to ensure they didn't increase Pinta's invasive mammal population, the Judas goats released on Pinta wore radio collars. Because goats are social, the island's new arrivals sought out friends, unwittingly leading hunters to the island's hold-outs.

If you're anything like me, learning about Judas goats (and animal eradications in general) leaves you feeling uneasy. Why should invasive animals have to pay the price of human error? It's not their fault they ended up in places where they don't belong, and it's not like they *intentionally* harm the native fauna and flora of their adoptive homes. There is no easy way to grapple with thorny issues like these, but how you feel about eradication – as a concept and a practice – depends, I suspect, on where you sit on the animal welfare–conservation spectrum. Do you believe in animal welfare regardless of the environmental cost or is conservation your primary concern? If you fall into the latter camp, as I do, you are open to accepting the loss of animal life in the name of conservation.

Philosophical consideration aside, Judas goats turned out to be a powerful tool in the eradication toolkit, and by 2003 Pinta was officially goat-free. Before long, national park employees and locals scaled up eradication, eventually removing 80,000 goats from Santiago Island in a span of 52 months. Donkeys and pigs were also targeted as part of Project Isabela. Although Pinta was ready for tortoises to return, conservationists weren't about to send Lonesome George home on his own, particularly since they still hoped he would mate, either with a yet-discovered Pinta tortoise or a related species.

The recovery of Pinta – both island and tortoise – had reached a standstill. Lonesome George was alive but not breeding, and although the island had been cleared of invasive animals, it needed giant tortoises to make a full recovery. At least Pinta had a chance, albeit a longshot, but what about Floreana and Santa Fé, both of which lost their tortoises in the mid 1800s? Without keystone herbivores, their ecosystems would never fully recover, and the more time passed, the further ecosystems drifted from their original state. When large herbivores, whether elephants or giant tortoises, are lost, plant communities become less rich, less varied. In turn, mutualistic relationships between insects and plants suffer, as do rates of nutrient cycling.

Scientists could have easily given up on Floreana, Santa Fé and Pinta, instead focusing their energy on islands containing tortoises, but conservation groups and the national park directorate were committed to restoring the Galápagos as fully as possible. Besides, they had another card to play.

As rewilding came on to the scene in the 1990s, a related field of science began to blossom. It involves what are called 'ecological replacements'. The concept is straightforward: when a species goes extinct, it leaves an ecological void that can be filled by introducing a similar species as a replacement. Simple in theory, the use of ecological replacements is complex – and controversial. Not only does the practice require playing God, but unforeseeable consequences may harm the replacement species as well as the ecosystem it enters. Questions about its efficacy also remain. Not only are replacement species different from the extinct ones they mimic, but some 'ecosystem disruptions have occurred in deep enough history to render functions irrevocably lost'. Put another way, even if scientists are confident in the abilities of the stand-in species, they need to be sure the ecosystem is prepared to welcome it (and vice versa). Climate change raises an additional question: can species adapted to past conditions survive now and into the future? In light of the risks involved, some conservationists oppose the use of ecological replacements. Others support its use as long as replacement projects are carefully considered and closely monitored.

A third camp led by risk-takers is convinced that ecological replacements are the best (and sometimes only) way to revive long-imbalanced ecosystems. Some of their proposals are downright radical. One involves translocating cheetahs and lions to North America to revive roles played by extinct big cats. An even bolder idea centres on returning woolly mammoths to the Siberian steppe using DNA editing and Asian elephant surrogates. Most of these ideas fall on deaf ears, but some have made it past the planning phase, as evidenced by India's decision to introduce African cheetahs as stand-ins for the country's former population of Asiatic cheetahs. The plan has plenty of critics, among them Indian conservationist K. Ullas Karanth, who says the initiative is doomed to fail. In his opinion, cheetahs will only survive Indian introductions if placed in areas that are free of dogs, people, leopards and tigers (in other words, a cheetah-friendly Shangri-La).

The more closely related the replacement species is to its extinct counterpart, the better. In the case of Galápagos tortoises, all 15 of the named species, 13 of which survive today, can be traced back to a common ancestor. You may recall that when Darwin visited the archipelago in 1835, he was told that tortoises of different islands featured unique sizes and shell shapes. Morphological differences can indicate genetic differences, but establishing the taxonomy of Galápagos tortoises proved a mighty challenge. For decades, scientists debated whether distinctive tortoises were separate species or simply subspecies. As the authors of a 1999 scientific paper pointed out, 'Perhaps the most enduring debate in reptile systematics has involved the giant Galápagos tortoises'.[9] Since then, the majority of experts have come to believe that the 15 distinct tortoises are in fact separate species, but the debate rages on.

Risks aside, many rewilding experts are pushing to make the restoration of large herbivores (and their ecological functions) a conservation priority. For several reasons, giant tortoises have been hailed as 'model taxon' for ecological replacement rewilding: the species they replace went extinct relatively recently; the risk of disease outbreak associated with giant tortoise reintroductions is low; and their slow pace and large size mean that individuals can be easily contained (and moved back to captive facilities if plans go awry).

It should come as no surprise, then, that one of the first-ever ecological replacement projects involved giant tortoises. The project, which began in the early 2000s and is ongoing, involves the Aldabra giant tortoise. Native to the Seychelles, the domed reptile is closely related to five species that went extinct in the nearby Mascarene Islands of Mauritius, Rodrigues and Réunion. Like the giant tortoises of Floreana and Santa Fé, the Mascarene species disappeared during the mid 1800s due to overharvesting by sailors and problems associated with the foreign species they introduced. Because they were the islands' dominant herbivores, their loss set off a cascade of negative ecological impacts.

Mauritius, land of the dodo, has experienced 'one of the saddest tales of island extinction'.[10] The country has lost 95 per cent of its native vegetation and 46 per cent of its native vertebrates,[11] but naturalists and scientists have spent decades battling to save what remains. One of the island's best-known advocates was Gerald Durrell, who founded the Durrell Wildlife Conservation Trust, helped establish the

Mauritian Wildlife Foundation, and authored dozens of books, one of which inspired a popular TV series, *My Family and Other Animals*. His 1977 book on Mauritius, *Golden Bats and Pink Pigeons*, features fascinating anecdotes and sparkling prose, 'Mauritius gleams in a million tropical greens, from the greens of dragon wing and emerald, to delicate dawn greens and the creamy greens of bamboo shoot.'

Twelve years before the book came out, Mauritius converted one of its small islands – Île aux Aigrettes – into a nature reserve. Its 27 hectares (67 acres) contain one of the last of the country's coastal forests, a formerly widespread habitat type where giant tortoises were common. More importantly, much of the island's flora has been restored, and its shores lack introduced animals capable of preying on tortoise hatchlings, making it an ideal reintroduction site.

Welsh conservationist Carl Jones has worked in Mauritius since 1979 and is best known for helping critically endangered birds like the Mauritius kestrel avoid extinction. Reflecting on Jones's achievements, the chair of the IUCN Species Survival Commission said, 'I know of no other conservationist who has directly saved so many species from extinction.' In addition to saving rare birds and reptiles, Jones pioneered the use of ecological replacements. 'When he started talking about using Aldabra tortoises to repopulate Mauritius in the 1980s, people thought he was nuts,' says colleague Vikash Tatayah, 'but they eventually came around.' In the mid 1990s, the Mauritian Wildlife Foundation gave ecological replacement rewilding a go when it released four Aldabra tortoises into a small enclosure on Île aux Aigrettes.

Although Aldabra tortoises are native to the Seychelles, a number of them have been kept in private collections in Mauritius since the late eighteenth century, so the NGO didn't have to look far when sourcing animals. Darwin, it should be pointed out, played a part in early transfers of Aldabra tortoises to Mauritius. In 2002, the four founders were moved to a bigger enclosure where they joined 12 additional Aldabra tortoises. They began breeding that year and were set free on the island the following year. In the project's early days, Vikash and his colleagues removed hatchlings and placed them in a captive breeding centre where they were 'head-started' for several years (poaching, particularly of young tortoises, remains a threat). Once ready for life in the wild, they began repopulating additional islands that had been cleared of invasive species. Of them, Round Island contains special significance since giant tortoises native to Mauritius were last seen there – in 1844. A nature reserve since 1957, the island's shores were cleared of pests (rabbits) with help from New Zealand's Don Merton, whom we met in Chapter Three.

Since the project's inception, scientists have closely monitored relationships between released tortoises and island vegetation, finding that tortoises are consuming (and thus dispersing) native seeds. They are simultaneously helping to control the spread of invasive plants by feasting on them, but some areas containing invasive vegetation are too rugged for tortoises to reach, so conservationists have to manually weed them. Hailed as a success, the project shows no signs of slowing down. Its greatest achievement arguably lies in its influence: by demonstrating that ecological replacements can enhance restoration while offering sanctuary to endangered species, it laid the foundation for future initiatives in other parts of the world.

When Aldabra tortoises began repairing Mauritian habitats, conservationists in the Galápagos were preparing to trial a similar approach. The year was 2005, and Lonesome George was approximately 95 years old. Not young, nor geriatric, the weathered reptile was no closer to reproducing. Still lonely, George had been given a new, equally depressing title: 'the rarest creature in the world'. I never had the chance to see him alive, but having met Sudan, I can imagine how visitors to the Santa Cruz Breeding Center felt when they laid eyes on him. The emotional cocktail of honour, hope, sadness and guilt that washed over them is, I'm convinced, reserved for witnessing extinction.

Hope for the Pinta tortoise was slowly frittering away when another discovery dramatically altered the course of Galápagos tortoise rewilding. It occurred in New Haven, Connecticut, in a Yale University lab run by Adalgisa 'Gisella' Caccone, an evolutionary biologist who has worked in the Galápagos since 1994. For several years, Gisella had been sitting on mysterious blood samples from tortoises found on one of Isabela Island's volcanoes, Wolf. She realised they were special in 2001 when she compared them with samples taken from Wolf's native giant tortoise, *Chelonoidis becki*. They were distinct, so much so that Gisella felt like she was looking at alien DNA. She finally solved the mystery in 2005 when she had the opportunity to compare the unknown DNA with genetic material obtained from two museum specimens – one of the extinct Floreana tortoise, the other of the nearly extinct Pinta tortoise (Gisella's analysis also made use of Lonesome George's DNA). She was over the moon when some of the 'alien' Wolf Volcano samples matched those taken from deceased tortoises. The results proved that Pinta and Floreana DNA were preserved not just in museum specimens, but in living, breathing, seed-spreading tortoises.

Armed with the knowledge that George wasn't completely alone and Floreana tortoises weren't fully extinct, the national park directorate organised another expedition to Wolf Volcano in 2008 to look for additional tortoises containing mixed ancestry. One of six volcanos on Isabela Island, Wolf is a bowl-shaped 'shield volcano' that reaches 1,707m (5,600ft) and is the tallest point on the archipelago. Still active, parts of the volcano are incredibly steep, and the surrounding area is comprised of variable terrain that would, I suspect, have been too 'fatiguing' for Darwin. Tricky logistics aside, inaccessible places like these are prized by conservationists seeking unspoiled nature or concealed wildlife.

Biologist Wacho Tapia led the 50-person expedition. The Galápagos native manages the Giant Tortoise Restoration Initiative (GTRI), a collaborative effort launched in 2014 by the Galápagos National Park Directorate and Galápagos Conservancy. Wolf Volcano is home to an estimated 8,000 giant tortoises, so Wacho and his crew needed to focus on finding individuals with unusual shell shapes. In particular, they scoured the area for saddle-shaped shells because Wolf's native tortoise is domed (as a reminder, of the 15 Galápagos tortoise species, five are saddle-backed while 10 are domed). Where the saddle-backed varieties came from they weren't sure, but Wacho and co collected blood samples from 1,600 tortoises and shipped them off to Gisella's lab.

Gisella already had samples from every species, but that didn't mean she could provide instantaneous results. A year later, she was able to conclude that the saddle-backed tortoises spotted by Wacho were a mix between Wolf Volcano's native species and tortoises from other islands. Of the sampled hybrids, 17 had Pinta ancestry and 80 had Floreana genes (others were part-Española). Because the Wolf Volcano population contained a small number of hybrids 'in an otherwise homogenous population', Gisella believes that hybridisation occurred relatively recently (recent enough that pure-bred Pinta or Floreana tortoises *may* be still be alive!). How did Floreana and Pinta's tortoises get to Isabela Island? Could individuals have floated or hitched rides on rafts like the archipelago's original tortoises did millions of years ago?

After considering the direction of ocean currents, conservationists concluded that natural migrations were extremely unlikely. Besides, old ship logs show that vessels frequently passed by Wolf Volcano while harbouring in Banks Bay, which is located at the volcano's western base. As we learned earlier, mariners tossed live tortoises overboard when a ship's load needed lightening, so Floreana and Pinta tortoises must have survived these drops, reached the shore, and carried on living on or near the volcano. 'The irony is that these species have a second chance for the same reason they died out – their handling by mariners some 200 years ago,' remarked Gisella.

Although 50 people participated in the 2008 expedition, further expeditions were required. Suspected hybrids, for instance, were left on the island after being sampled and tagged. Complex and limited resources meant the team could only accomplish so much. In the meantime, Wacho and his counterparts at the national park directorate decided to begin repopulating Pinta Island with ecological replacements. In May 2010, they translocated 39 adult tortoises that contained mixed ancestry to the island. Pinta's native tortoise is saddle-backed, but translocated hybrids possessed a mix of domed and saddle-shaped shells. It was an intentional decision, an experiment, to evaluate how hybrids would adapt to (and shape) new environments. Importantly, scientists sterilised each of the released tortoises before placing them on Pinta. That way, if George or a hybrid discovered on Wolf Volcano later produced offspring, conservationists could return pure-bred tortoises to Pinta without worrying about a muddied gene pool.

Responsible conservation requires many steps – from assessments and planning to interventions and monitoring. In many rewilding projects that hinge on species reintroductions, population growth is the main metric for evaluating success. We saw this in Chapter Two. In Chad, Justin Chuven and his colleagues hope to see the scimitar-horned oryx herd grow to 500 individuals. However, as biologists pointed out in a 2014 paper, 'this criterion is inadequate when species reintroduction is undertaken to restore ecological functions and interactions', as in the Galápagos. In other words, population growth does not guarantee healthier ecosystems. Because projects like the Giant Tortoise Restoration Initiative are less common than traditional species reintroductions, scientists running them are charting new territory. Española Island – the first in the Galápagos to regain giant tortoises after losing them – was a critical testing ground.

First reintroduced in 1975, captive-born Española tortoises fared well in the wild (more than half of all released tortoises were still alive in 2007, a percentage scientists consider high, and 'significant' reproduction had helped the island's population rapidly grow to more than 800). By 2014, more than 1,000 tortoises were on the island; and the population has since risen to around 2,300. Scientists used conservation modelling to estimate the group's future extinction risk, concluding that, with or without additional reintroductions, the population has a very low chance (1 per cent) of going extinct. In terms of population growth and stability, the experiment worked, but when it came to habitat restoration and ecological functioning, the picture was less rosy.

As on other Galápagos islands, Española's vegetation experienced dramatic changes during recent centuries as invasive animals, namely goats, dominated landscapes formerly occupied by giant tortoises. As a result, woody plants became widespread and cacti communities diminished. Scientists hoped that giant tortoises would restore the natural balance and resurrect a critical relationship that exists between Española tortoises and cacti. After all, tortoises are ecosystem engineers. What scientists found instead was that reintroduced tortoises were struggling to make headway on their own. Impeded by woody vegetation, they stuck to their preferred habitat – areas that contain high densities of adult cactus and limited woody plants – despite the fact that such habitats are currently extremely limited on Española.

Tortoises weren't fanning out as anticipated and couldn't, therefore, engineer plant communities *throughout* the island. It became clear that people would need to lend a helping hand by removing woody vegetation and planting and seeding cacti. Unfortunately, woody plants have proven extremely difficult to control. 'With woody vegetation, we have kind of thrown our hands up in the air,' laments James Gibbs, a biologist and professor who has collaborated with Wacho and Gisella for decades. 'There is no way to manually handle the problem, and controlled burns aren't an option since fire is not a natural process in the Galápagos.' Big cacti have rebounded, however, thanks to efforts launched in 2017 as part of the Galápagos Verde 2050 project (an initiative focused on sustainable agriculture and ecological restoration). Restoration focuses on a critically endangered cactus species – *Opuntia megasperma* – that has a mutualistic relationship with Española tortoises. Put simply, the two need each other to survive.

In addition to underscoring that population growth is not the single nor most important marker of success, Española's rewilding illuminates another critical lesson about giant tortoise-led restoration. Such initiatives typically pursue two goals – population growth and ecological restoration – but these goals are unlikely to be met simultaneously. Even though giant tortoises take a long time to reproduce, population viability will likely be reached before 'ecosystem-level effects' manifest. That's because *more* animals are needed to achieve ecological restoration than are required for 'population persistence alone'.[12]

Española's rewilding taught conservationists a great deal about kickstarting ecological restoration using *native* species, but a paramount question remained: would the lessons they learned apply to ecological replacement projects, or would islands welcoming non-native tortoises (or hybrids) face challenges of their own? That became the million-dollar question.

To begin answering it, scientists turned their attention to Pinta since it was the first Galápagos island to receive hybrid tortoises in lieu of native ones. Following the May 2010 release of 39 sterilised hybrids – all of which wore tracking devices – conservation managers remained on the island and spent two months monitoring the released reptiles. Conservationists returned a year later to see how tortoise movement and behaviour had changed, if at all, during the tortoises' first 12 months on Pinta. What they found was unsurprising: domed tortoises, which generally live at higher elevations, had moved upland even though doing so meant abandoning cactus-rich areas. They occasionally consumed fallen cactus pads and fruits when migrating downwards before nesting, but otherwise remained in areas devoid of cactus. Saddle-backs, meanwhile, moved to new areas but remained at points of mid-elevation where they were able to feed on their food of choice, cactus. By dispersing cacti seeds, saddle-backed hybrids fulfilled one of their key duties as ecosystem engineers. They also utilised more of the island than their domed counterparts, which further increased their environmental impact.

Because the *Opuntia* cactus is a keystone species in arid portions of the Galápagos, restoring it (and its interactions with giant tortoises) is paramount. On top of providing food and water, *Opuntia* offers shelter and shade to a range of animals. In light of its importance, scientists concluded that domed tortoises were not a suitable replacement for Pinta's native species. By contrast, saddle-backed hybrids filled the ecological niche crafted by Lonesome George's ancestors. Notably, the saddle-backed hybrids released on Pinta were not closely related to the island's native species, suggesting that *morphology* may be more important than *relatedness.*

For years, Wacho patiently waited for the chance to return to Wolf Volcano. Not only did he hope to find additional tortoises containing Pinta and Floreana ancestry, he wanted to probe more of the area, a large portion of which remains unexplored. Perhaps Wacho would discover a pure-bred tortoise from a presumed-extinct 'Lazarus' species; or maybe he'd come across a brand-new plant or animal. At the very least, he was likely to locate more hybrids, some of which could contain higher degrees of Pinta and Floreana ancestry, making them even more valuable to captive breeding programmes than previously tagged individuals.

Finally, in December 2014, Wacho got his wish. Along with two park rangers, he and James Gibbs travelled to the island and set up camp near the coast. Every morning began with a 4-km (2.5-mile) trek across harsh terrain towards the edges of the caldera-shaped volcano. They worked in teams of two, combing areas where hybrid tortoises were located in 2008. Day one was a bit of a bust, but although James and Wacho only encountered several tortoises, they were delighted to see that northern Isabela's vegetation was flourishing in the wake of goat eradication. 'In 2008, only a couple of years after eradication, the area remained completely open and poorly vegetated,' wrote Wacho in a blog post.[13] By late 2014, it was transformed. A dense forest enveloped the volcano's slopes.

The following day, the foursome ventured to the highlands. Again, they failed to find tortoises, but their luck changed when Wacho noticed fresh tortoise tracks during a light rainfall. Tortoises, it seemed, had recently descended from the highlands. Following the tracks, the team eventually came to a gully filled with shallow pools of rainwater. All along them were tortoises – dozens of them – quietly drinking in the rain. Because Wacho and James were losing light, they quickly took blood samples from tortoises containing full or semi-saddle-backed shells before returning to camp. When they awoke the next morning, they rushed back to the gulley and continued sampling (and tagging) tortoises with obvious Floreana or Pinta ancestry, though Gisella's lab would need to confirm the men's suspicions.

As before, suspected hybrids were left on the island when the expedition came to a close. When I read about Wolf Volcano expeditions, I couldn't help wondering why scientists left hybrid giant tortoises behind given their rarity and value. Pinta, at least, has sterilised tortoises walking its shores, but Floreana hasn't had tortoises since the 1850s. Given the urgent need for restoration that only giant tortoises can provide – and the high stakes associated with preserving the DNA of extinct species – shouldn't James and Wacho have loaded hybrids on their backs, as whalers did before them, and taken them to Santa Cruz? I put this to James when we spoke last summer.

He explained that scientists prefer leaving potential hybrids in the wild until DNA tests reveal their ancestry. Unfortunately, export permits and the like mean that results sometimes take a year. 'In an ideal world, we'd have real-time PCR on board the ship with park guards sending high priority samples for overnight genetic confirmation, but the tech required to rapidly distinguish hybrids is not yet available,' says James. Instead, tortoises are tagged, sampled, and sent on their merry way.

Given the long lifespan of giant tortoises, another year or two on Isabela Island should not impede a tortoise's future breeding and reintroduction prospects. Wolf Volcano, however, is active, and liable to spew lava at any time. I asked James about an eruption that occurred in 2015 and whether future episodes might jeopardise Isabela's rare tortoises. 'The volcanism issue,' James replied, 'is very real.' He and Wacho were on Isabela during the 2015 episode, an experience James called 'mildly terrifying'. Fortunately, lava flowed in such a way that it averted the areas where tortoises and iguanas live, but if a future eruption sends lava flowing to the north or west, it will 'be a disaster,' emphasises James.

Not only does Wolf house upwards of 8,000 tortoises, but it contains the rare and critically endangered pink land iguana. Discovered in 1986, the land iguana was deemed a distinct species in 2009. Only 200 remain, and the entire population lives on the volcano's slopes, so a single eruption could erase the species. Plans are underway to translocate pink land iguanas to Santa Cruz for captive breeding, and the species received a massive boost in 2021 when a new Galápagos rewilding initiative made their recovery a priority. With a $43-million pledge and numerous partners, including the Leonardo DiCaprio Foundation, the endeavour adds momentum to the archipelago's long-running rewilding movement.

By November 2015, Gisella was able to confirm that numerous tortoises sampled in May contained Pinta and Floreana ancestry, so the national park organised another expedition to Wolf Volcano to locate and retrieve hybrids. Large nets dangling from helicopters carried tortoises from the volcano to a nearby boat. The team was only able to remove 32 tortoises – the most, Gisella said, that a boat could safely transport to the Santa Cruz Breeding Center. Of them, 19 contained Floreana ancestry and two were part-Pinta. One, a male, is said to be a spitting image of Lonesome George, who died of natural causes three years earlier, in 2012.

His death shocked the world. George could have carried on for another hundred years, and for many of us his passing marked the first time a species went extinct on our watch. Although George seemed uninterested in breeding and lacked a pure Pinta mate, his existence sparked optimism for his species. George was a link to the past and a bridge to the future. He may not have remembered people removing the last of his kind from Pinta's shores, but he was there when it happened. When I read about his passing, I tried to imagine the precise moment when he became the last tortoise on Pinta. How old was he when it happened? Did he lay eyes on the people who took his last-surviving relatives away, or was he on another part of the island, safely concealed from view, when the retrieval occurred?

I gazed upon George's taxidermised body while visiting the Santa Cruz Breeding Center in 2019, and boy was he lonesome. Encased in a large box of wood and glass, George is kept in a dark and isolated portion of the facility to prevent damage caused by sunlight passing through nearby windows. For several contemplative minutes, I stood by George. I even gave myself permission to anthropomorphise his life – a cardinal sin, according to most scientists. Did George experience anything akin to loneliness? After years on his own, was he, like me after months of lockdown, uncomfortable reuniting with other animals? Even now, I can't shake the feeling that George might have preferred to live out his final years alone on Pinta. On the other hand, I can't imagine any responsible conservationist electing to leave him behind. Like it or not, George had a part to play in helping to save his species.

I looked at George one last time and walked away. Outside, I rejoined my tour group, and soon we were climbing aboard a bus bound for the highlands. There, we had the chance to see Santa Cruz tortoises roaming free. I took photos of one resting in the shade and another making its way to the edge of a small pond. Seeing the large reptiles up close allowed me to understand the awe Darwin felt during his first encounter. I can't begin to imagine how the Dominican friar who first encountered the animals in 1535 must have felt at the sight of them. Did he, like Darwin, leap on one's back?

Visitors aren't able to observe the hybrid tortoises retrieved from Wolf Volcano – their rarity keeps them off limits – but considerable progress has been made since they were removed from Isabela in 2015. Two years later, scientists launched a Floreana tortoise breeding programme using Wolf hybrids and tortoises with Floreana ancestry that were already residing at the breeding centre. To maintain genetic diversity, all of the 23 'genetically important individuals' are participating in the programme regardless of their mitochondrial lineage, which means that releases may begin before complete genome recovery occurs. That's *if* the national park directorate decides to prioritise reintroductions above genetic purity. If, on the other hand, the national park wants to 'breed back' pure (or nearly pure) Floreana tortoises, reintroductions will have to wait. By Gisella's calculations, 90 per cent of the Floreana genome could be recovered in four to five generations of captive breeding, but given the tortoise's long lifespan, that would take several hundred years.

For his part, James is intrigued by the idea of sending hybrids to Floreana. "They have shreds of the original genome,' he tells me, 'but some of the differentiation in their genetic makeup may help them in the future.' That's because Floreana has changed since its giant tortoise went extinct, and climate change will further alter the island. James emphasises that 'a diverse set of traits could help the tortoises re-adapt and reshape their lineage'. On the other hand, conservationists could transfer some of the purebred Española tortoises hatched with Diego's help, but they suffer from low genetic diversity, he warns. Besides, the national park directorate is unlikely to have a viable alternative, and islands missing giant tortoises *need* them back. Gisella agrees. She believes that putting hybrids back and letting selection run its course is a way to 'let the island choose' its future.

On 2 August 2021, 30 rangers and scientists embarked on a highly anticipated 10-day expedition to Wolf Volcano. James and Wacho planned the expedition but were unable to attend: work obligations held them back. Backpacks, blood collection equipment and tents were crammed into two small boats bound for the remote volcano.

Although biologists and rangers spent some of their time on Wolf looking for non-native giant tortoises like the hybrids tagged in 2008 and 2014, they primarily focused on pink land iguanas as they conducted the first-ever comprehensive census of the species. For days, conservationists scoured the volcano's slopes looking for critically endangered animals. To estimate the volcano's pink-hued iguana population in a short period of time, researchers utilised what's called the 'mark-recapture method' whereby animals are trapped, given a harmless-but-visible mark on their body, and released. Then, days later, traps are set again. Researchers arrive at an estimated population size by multiplying the number of animals trapped and tagged the first time by the number of animals trapped the second time. This figure is then divided by the number of tagged animals that turned up in the second batch.

Using this mathematical formula, researchers believe that approximately 211 pink land iguanas exist today. While low, the figure aligns with previous estimates, however researchers were devastated by another of their findings: none of the iguanas they came across were juveniles. All were adults. The last juvenile was seen in 2014, suggesting that the population may not be growing at all. Feral cats and rats could be to blame. Before leaving Wolf, researchers placed motion-activated cameras around the volcano's summit. That way, they can better understand the threats facing the rare lizard and, in turn, better prepare for its conservation. Several days later, one of the remote cameras broadcast a rodent hanging around a pink iguana nest, suggesting that rodents may prey on iguana eggs and perhaps even hatchlings.

Reporting on the expedition, the Galápagos Conservancy insists there is hope for pink iguanas. The NGO pointed to the fact that a wide range of stakeholders have come together to craft a conservation management plan, the first for the species. Captive breeding will be central to it, but according to James, conservationists first need to study the iguana's biology and assess the threats facing it. In the meantime, scientists hold out hope for the possibility of discovering pure-bred, presumed-extinct species.

Their optimism is warranted.

In 2019, during an expedition led by Wacho, a national park ranger spotted a female tortoise on Fernandina, an island believed to have lost its native tortoise, *Chelonoidis phantasticus*, 112 years ago. Was she the next Lonesome George or a tortoise displaced from another Galápagos island? Wacho and his team moved the lone female, who is believed to be around 100 years old, to the Santa Cruz Breeding Center as there was no guarantee they would find her again. Blood samples, meanwhile, went to Gisella, who checked them against the only *Chelonoidis phantasticus* specimen on record – a male retrieved during the California Academy of Science's 1905–6 expedition.

Finally, in May 2021, Gisella and her colleagues shared their preliminary findings. The female, now called 'Fernanda', is related to the male collected more than a century earlier, and she may not be the last of her kind. Park rangers have come across additional tortoise tracks on Fernandina and plan to launch a series of expeditions to search for relatives. Like Lonesome George and the newly named pink land iguana, Fernanda reminds us that nature has yet to reveal its every secret. The Galápagos helped Darwin unravel what he called the 'mystery of mysteries – the first appearance of new beings on this earth', but his work is not done. At its core, rewilding seeks to restore and rebalance, but in the Galápagos and beyond rewilding offers something more: a cause for exploration and the chance of discovery.

CHAPTER NINE

Connecting the Cores

A powerful image pulled me into the story of China's wandering elephants. I suspect you came across the picture, too, given its sweeping distribution. Taken from a drone on 7 June 2021 in the country's Yunnan province, the photo depicts a small herd sleeping in a pyramid-shaped huddle. A calf resting on its right side is neatly positioned between two larger elephants. The smallest piece in a jigsaw puzzle is closely guarded – even during sleep.

The elephants are part of a herd that left their home in a wildlife reserve during April 2020. By June 2021, they had travelled around 500km (310 miles) while making their way north from the south-west of Yunnan to the province's capital, Kunming. Mid-summer, they shifted course and began moving in a south-westerly direction. Perhaps they were heading home, or maybe they're destined for a life on the road. No one knows where the elephants are going or why they traded the safety of a protected area in favour of human-dominated landscapes, but scientists have several theories.

Both Asian and African elephant herds are led by matriarchs appointed based on age and experience; a young or inexperienced leader could explain the haphazard peregrinations of China's famous herd. Other scientists see methods in the madness – or at least an explanation for it. Although Asian elephants have defined home ranges, they are known to deviate from their homes when pinched by external forces like drought or habitat loss. As historic habitats shrink in size and diminish in quality, elephants are forced to search for greener pastures. Could it be that China's mobile herd had nowhere else to go?

Nilanga Jayasinghe, who manages Asian species conservation for the WWF, points out that significant habitat loss has occurred within the herd's home range during recent decades. She suspects the herd got lost while seeking new territory. The Chinese government poured significant resources into keeping the elephants – and the people they encountered – safe during the herd's unpredictable journey. Followed by 20 drones and a 360-person task force, the herd was monitored 24/7. Social media broadcast the elephants' antics, and the world tuned in. People saw when a calf needed a mother's helping hand to climb out of an irrigation ditch and watched, transfixed, as five elephants casually strolled through a car dealership.

Although no people (nor elephants) were harmed during the months-long saga, the herd consumed and trampled crops, knocked down fences and raided shops, causing an estimated $1 million in economic damage. The task force deployed numerous tactics to prevent further destruction and correct the herd's course. Food was dropped in strategic locations to lure the elephants away from villages and turn them around. Roads were blocked, and villagers were periodically evacuated when the herd got too close. And, yet, the elephants kept moving.

Given what we know about translocation as a conservation tool, you may be wondering why the Chinese authorities didn't forcibly return the meandering elephants to their home base. In fact, they did relocate one member of the herd, a male that split off from the group on 6 June 2021 – 15 months after the herd's odyssey began. After spending a month on his own, it became clear that the male was disinclined to rejoin the others, so authorities anesthetised him and transported him to the reserve he left in April 2020. It was a risky move.

While elephants have been shifted between areas – sometimes in large numbers and sometimes with great success – they are known to rebel against forced translocations and return to their preferred location even if it means trekking long and unfamiliar distances. Sri Lanka learned this lesson the hard way. For decades, the government moved 'problem elephants' (those involved in crop-raiding or other forms of human–wildlife conflict) into fenced protected areas only to find that the majority attempted to return home.

In one case, 16 Sri Lankan elephants were translocated to several national parks. Within a year, 14 exited the parks where they were released (the quickest to leave departed a day after arriving). In some cases, fences were knocked down as elephants broke free. The other two died near the protected areas where conservationists let them loose, shot by people for crop-raiding. Translocation – and the adjustments that follow – can be tremendously stressful for intelligent and social animals like elephants. Scientists monitoring the project concluded that the 14 escapees did not leave in search of greener pastures since the release sites contained abundant natural resources and potential female mates. Likewise, conflict between elephants was not a cause for escape since no fighting occurred between translocated elephants and established ones.

'Asian elephants have well-defined home ranges with high fidelity,' wrote scientists in 2012, 'and it is likely that translocated elephants left the parks where they were released in attempts to return "home".'[1] Not only were experts unsurprised when elephants fled release sites, they were far from astonished when individuals prone to crop-raiding continued stealing from farmers.

Crop-raiding, in their experience, stems from preference versus necessity.

In Sri Lanka, 65 per cent of wild elephants live outside of protected areas. Attempts to move elephants into national parks – whether by corralling or translocating them – have failed, causing scientists like Pruthu Fernando to push for a different approach. Rather than fencing elephants *in*, he suggests fencing them *out*. Doing so involves placing barriers, like electric fences, around villages. Although locals initially bristled at the idea, a successful trial in a village called Galewewa validated the method. To date, no people or elephants have died, and nocturnal crop-raids are few and far between. Forty villages currently participate in the programme, and many more are poised to join now that the country's National Policy for Elephant Conservation and Management has formally adopted Pruthu's solution.

A pioneer in Sri Lanka, Pruthu didn't invent the fence-out method. In Africa, conservation organisations like Save the Elephants have long utilised barriers to deter elephants from entering villages and agricultural areas. Some are conventional electric fences; others feature natural deterrents like beehives and chilli peppers, both of which are repellent to elephants. These affordable solutions have an added bonus: they generate additional revenue streams for communities. They aren't foolproof, of course.

Elephants can – and will – go wherever they please, as evidenced by China's big-footed explorers. But if experts are able to understand what drives elephant movement – whether an ingrained, ancestral route or a dislike for spicy chillies – they can make management decisions that have a better chance of conserving wildlife while protecting people.

When I first read about China's roving elephants, my brain was tied in knots. From the moment I began crafting an outline for this book, I was determined to write a chapter on rewilding's third 'C': corridors. Many chapters touch on the importance of connectivity between cores, but I wanted – needed – to write a chapter where corridors are central to rewilding. Not only that, I hoped to identify a project involving transboundary conservation *and* species recolonisation. I could picture it: a new wildlife corridor spanning an international border. Because of it, fragmented groups would reunite for the first time in years in what would have made for one hell of a happy ending.

Of all my ideas, it was the most ambitious, but I was confident it could be done. I cast a wide net and was, for a time, planning to write about big cat conservation along the border of China and Russia, but only one objective expert source would speak with me. He acknowledged that, although tigers and leopards are moving between the two countries with greater frequency, the countries' collaborative efforts aren't rewilding per se since big cats were never prevented from crossing country lines. I also had a sneaking suspicion that when I interviewed representatives from each country's government, only superficial statements would be shared with me – language akin to, 'We are delighted to have this opportunity to work together to help big cats.' Since the pandemic prevented me from conducting in-person reporting, we were in for a real snooze-fest.

So, I set the idea – and my compulsion to detail an example of international connectivity – aside. Relieved of this requirement, I zoomed in on the Santa Monica Mountains of California. As many locals know, a male mountain lion known as P-22 successfully crossed two large highways to carve out territory in the middle of

Los Angeles. Cars kill between 75 and 100 mountain lions in California each year, making P-22's crossing of two mega-highways genuinely astounding. He occasionally pops up beneath people's decks, giving them the scare of a lifetime, but P-22 has become as famous as Lonesome George, perhaps more so given that he, like all big cats, is a symbol of raw power. It doesn't hurt that someone branded him the 'Brad Pitt of mountain lions'.

His prospects aren't as grim as George's were – mountain lions, which are also called cougars and pumas, are found in 28 countries – but as the only big cat in Los Angeles, P-22 is mate-less. If he dies before reuniting with the mountain lions living on the other side of the highways he courageously crossed, LA's one-cat population will go extinct. To prevent this from happening (and protect the region's mountain lion population from further inbreeding), conservationists hope to construct a 61m (200ft) wildlife crossing across the 101 interstate. The Liberty Canyon wildlife crossing will cost an estimated $85 million, $72 million of which had been raised as of February 2022. As tempted as I was to detail the crossing in this chapter, the project wasn't far enough along for me to do so*.

I was back at square one and running out of time. There were dozens if not hundreds of connectivity-oriented rewilding projects to uncover, but something drew me to Asian elephants – and India.

When it comes to conservation, you can't talk about one without the other. For starters, India contains around 60 per cent of the world's Asian elephant population, making it the species' last stronghold. The same goes for tigers. Indians have lived alongside elephants for thousands of years.

---

* Construction is due to begin on 22 April 2022, in tandem with Earth Day, and the bridge is slated to be completed in early 2025.

It's no surprise, then, that elephants feature prominently in Indian culture. For Hindus, in particular, elephants are sacred since they are considered a living incarnation of a celebrated Indian God, Ganesha. Even among the 20-odd per cent of Indians who don't identify as Hindu, elephants often possess cultural or traditional significance.

In many parts of the country, the word 'elephant' is the same as 'mother' or 'father' in one's local tongue, says conservationist Divya Vasudev. Does this mean people love and revere elephants, or fear and loathe them? As with human relationships, connections between Indians and elephant are 'complex', explains Divya. People may admire elephants, but fear is never far away. Among Indians, there is a certain degree of acceptance that comes from a long history of sharing space with wild, sometimes dangerous, animals, but tolerance has its limits.

Human–wildlife conflict is always a risk in places where people live in close proximity to destructive or dangerous animals. Elephants are both: not only do they raid crops, uproot trees and damage structures, they directly threaten human life. Each year, upwards of 500 Indians are killed by elephants. In return, around 100 elephants die in what are known as 'retaliatory killings'. Some retaliations are downright disturbing. In 2017, a photographer captured a shocking photo of a mother and calf running across a road after catching fire. The elephants may not have been intentionally set alight, but people were undoubtedly behind the injuries (Indians are known to throw firecrackers at elephants when they come too close). The haunting image is a stark reminder of the price wild elephants pay for daring to exist in a human-dominated world.

Mystery shrouds the status of the Asian elephant (*Elephas maximus*). Survey techniques vary across countries, making reliable estimates hard to come by, and forest-dwelling animals are notoriously hard to count. Conservation organisations often say that between 30,000 and 50,000 Asian elephants survive in 13 countries, but these figures are nothing more than educated guesses, warn experts.

Amid guesswork is some certainty. Global populations have been in decline for centuries and continue to dwindle today. *Elephas maximus* has vanished from large portions of a historic range that once spanned 900 million hectares (2.2 billion acres). It likely went extinct in the westernmost reaches, which began in Iran, as early as 100 BC. As now, ivory made ancient elephants susceptible to hunting and poaching, but unlike African elephants, *Elephas maximus* has long been domesticated and put to work. Even today, elephant keepers, or mahouts, force trained elephants to haul timber. According to The Nature Conservancy, Myanmar possesses the largest captive elephant population in the world. There, around 5,500 timber elephants help the country clear its forests – a perverse task given the elephant's need for intact habitat.

Surviving groups of Asian elephants are highly fragmented, as are the landscapes that sustain them. Habitat loss is driving extinction across the world but is especially detrimental to elephants and other 'mega-herbivores'. Because of their size, large food intake and need for mobility, elephants require sizeable core areas. Severe habitat fragmentation means they are unlikely to enjoy a contiguous home range, hence the need for corridors. Traditionally defined as linear, often-narrow stretches of habitat, a number of contemporary scientists emphasise that a corridor's value lies in function, not form.

Connectivity is critical for several reasons. For starters, corridors facilitate gene flow. Populations limited to a particular zone may be able to survive in the short term, but over time genetic diversity will be lost unless newcomers magically appear. They are unlikely to do so – and translocating elephants is risky business, as we know from Sri Lanka. As a result, conservationists cannot actively manage genetics across groups like they can with animals like lions. As an aside, India's tiger population *can* be managed using translocation, but cores and corridors are the *preferred* conservation approach.

Mobility minimises the risk of overcrowding, a problem that has numerous deleterious effects, from animal mortality to habitat degradation and resource depletion. Corridors alleviate these pressures while allowing herds to follow ingrained migratory routes irrespective of human development. By *encouraging* elephants to utilise wild spaces versus land occupied by people, corridors reduce conflict.

Scientists believe that herds are loyal to specific, enduring routes and ranges, which is why China's roving elephants had them scratching their heads. African and Asian elephants have large home ranges, in part because they move around in pursuit of resources. When it comes to Asian elephants, home ranges are poorly understood, but there's no doubt about the fact that sizes vary, sometimes dramatically. A herd living in a national park near the Himalayan foothills was found to have a 3,000–hectare (7,413-acre) home range.[2] In southern India's Nilgiri Biosphere Reserve, researchers who monitored three clans found a 65,100–hectare (160,866-acre)[3] mean size. Size-wise, African elephant home ranges fluctuate even more dramatically. According to the IUCN African Elephant Specialist Group, home range sizes can be as small as 1,500 hectares (3,707 acres) and as big as 370,000 hectares

(914,289 acres).[4] All of which begs the (unanswerable) question: how much space did elephants utilise before we took their habitat away?

Located in the Western Ghats mountain range of southern India, the Nilgiri Biosphere Reserve sprawls across three states: Karnataka, Kerala and Tamil Nadu. It contains one of India's four remaining elephant populations; the other three are in the north-west, north-east and centre of the country. With an estimated 6,500 individuals, the southern population – which some call the Brahmagiri-Nilgiris-Eastern Ghats population – is the largest in India, perhaps Asia.

Calling the biodiversity hotspot a 'reserve' is a bit misleading because only 22 per cent of its 552,000 hectares (1.4 million acres) are strictly protected core areas. The term 'reserve', says the Wildlife Trust of India's Upasana Ganguly, denotes an area's ecological importance. The hotspot's cores are comprised of six national parks and three wildlife sanctuaries that form a rough semicircle. Although some elephants stick to defined home ranges, others move between portions of the biosphere reserve, whether following historic migratory routes or seeking something new – a mate, perhaps, or fresh territory. Adult males, on the whole, travel longer distances than female-led family groups.

In some parts of the Nilgiri Biosphere Reserve, elephants are able to hopscotch between national parks and reserves. In others, protected areas are divided by human-dominated landscapes, forcing elephants and people to cross paths. On the biosphere's western side are two large national parks – Nagarhole and Bandipur – that jointly span 151,640 hectares (374,711 acres). Their size and location make the parks a critical corridor for migrating elephants as well as a prime base for resident pachyderms. Together, Nagarhole and Bandipur contain India's largest population of elephants as well as hundreds of tigers. Precise elephant numbers are hard to come by, but experts believe that around 3,000 individuals are based in the parks, which represents approximately 11 per cent of India's total population (27,312 at last count in 2017).

For a suite of reasons, protecting this population is critically important. Like many of the animals at the centre of rewilding projects, elephants are a keystone species that exert above-average influence on their environments. They disperse seeds, open up dense vegetation and during dry seasons use their tusks to dig holes in the ground to create pools from which numerous animals drink. Elephants are also flagship and umbrella species. Flagships are iconic and loveable and regularly feature in our culture. Think of WWF's logo (the panda) or the many pub signs bearing lions. If we admire these animals enough to incorporate them into our culture, it stands to reason that we are concerned about their protection. Umbrella species serve a different function. They are typically wide-ranging animals with large habitat requirements. As a result, conservationists are able to protect numerous species while focusing on one, which simplifies the arduous task of conservation planning.

There's no disputing that India has reason to conserve its elephants, but what about rewilding? How does it factor into the equation? When the founders of the rewilding movement articulated the Three Cs, they pushed for large-scale initiatives. Yellowstone to Yukon – an effort to connect two vast wilderness areas separated by 3,200km (2,000 miles) – comes to mind. But just as the 'C' for carnivores expanded to include non-carnivorous keystone species, rewilding has experienced a relaxation of core and corridor targets. In Chapter Eleven, we'll take a look at micro-rewilding, but let's stick to India for now.

As of 2019, 1.37 billion people lived on the subcontinent, which represented 18 per cent of the global population. That number is predicted to reach 1.66 billion by 2050, and India is poised to overtake China as the world's most populous country by 2027.[5] Given these stats and what they suggest about India's already fragmented wilderness, it's safe to say that Indian conservationists won't launch their own version of Yellowstone to Yukon any time soon. Rewilders instead initiate smaller-scale projects while conservationists fight to maintain ground.

Rewilding 'cannot be replicated in India the way it was originally developed in the West', argues Bahar Dutt, the author of *Rewilding: India's Experiments in Saving Nature*.[6] Dutt hits the nail on the head in the introduction to her book, 'Different countries have borrowed the concept and created their own version of it, which begs the question – can the concept be applied to India, where biodiversity and humans live together like tossed salad, where protected areas don't exist on the same scale as in Africa or Europe?' Dutt says Western conceptions of rewilding have limited applicability to India but believes in building the case for rewilding in places 'where biodiversity thrives in a sea of humanity'.

I wholeheartedly agree – and not just because I support diverse rewilding approaches. If rewilding is relegated to large, remote wilderness areas, the majority of people will view the practice as something distant and fringe – if they connect with it at all. The resulting disconnect will threaten existing projects (if people can't engage with something, they are less likely to support it) while inhibiting new ones (if rewilding only occurs *out there*, people are less likely to pursue rewilding where they live). Furthermore, as an increasing number of people move to urban areas, rewilding in human-dominated landscapes will take on even greater importance.

In 1971, Ranjitsinh Jhala was serving his first term as India's director of wildlife preservation. Now in his 80s, the author and conservationist spent 1971 drafting India's Wildlife Protection Act, which passed both houses of parliament the following year. Among other things, the Act made elephants a protected species and put an end to a longstanding practice called 'khedda' in which elephants were captured and broken down until they submitted to domestication. For centuries, khedda was practised throughout India, but Ranjitsinh only knew of one permanent capture site: where the Kabini River meets Nagarhole and Bandipur (the two parks that together contain India's largest elephant population).

'Khedda was fine-tuned there,' says Ranjitsinh. 'With a valley and slopes on either side, the topography of the Kabini area lent itself to a permanent capture site.' International tourists paid top dollar to witness khedda, lured, at least in part, by the region's royal ties (the former rulers of Mysore owned a hunting lodge in Nagarhole). During the final khedda, which took place in late 1972 or early 1973, pandemonium broke out, recalls Ranjitsinh, and several elephants were shot. One witness remembers watching in horror as 'the drummers drove all the wild elephants into the water as the domesticated elephants surround[ed] them. They were then driven into an enclosure and were caught by ropes ... some used to fall into the pits and we could hear their cries; it was so sad'.[7]

Ending the cruel practice was an important step towards rewilding, both for elephants destined for captivity and populations as a whole. Trauma, for elephants, can leave deep scars that alter behaviour in multiple ways. Some herds experience social disarray after losing key members. Others avoid areas where traumatic events occurred, but in Karnataka, elephants continued visiting the Kabini River after years of khedda, perhaps because its resources are too rich to bypass.

As khedda became a practice of the past, another disruption to elephant mobility descended on the stretch of India where the Kabini River intersects Nagarhole and Bandipur. This time a dam and reservoir built along (and named for) the river bifurcated the two parks, which had been contiguous, thus threatening longstanding migratory routes. As we learned earlier, elephants typically have fixed home ranges that pass down through the generations. For elephants, geographic decisions boil down to habitat suitability, safety and access to resources and mates. Although the Kabini River naturally separates the parks, rivers do not impede elephant movement. A large reservoir was a different story: Ranjitsinh worried that it might act as a barrier.

By the time he learned about the project (in 1971), construction was already underway, making resistance futile. 'It was a *fait accompli*,' Ranjitsinh told me during a phone interview. In 1974, the Kabini Dam and Reservoir were complete. Beneath them lie the old khedda grounds whose shameful history has all but vanished from living memory. Although the reservoir supplies water to 22 villages, 14 settlements and parts of Bangalore, damming the river created a litany of problems while solving one. The reservoir submerged part of Nagarhole and swallowed villages whole. A number of communities were forced to relocate. For members of the forest-dwelling Jenu Kuruba tribe, this came as a double blow: the 1972 Wildlife

Protection Act restricted their access to the forest and now they had to uproot. In the coming decades, more indigenous communities would be asked (or compelled, depending on your point of view) to relocate in the name of conservation.

When it came to elephant mobility, some of Ranjitsinh's concerns were laid to rest when he witnessed herds swimming across the reservoir using their trunks as snorkels. Not all elephants are comfortable making the journey, however. Calves, for instance, are unable to swim until they are several months old, and small groups are less confident crossing, explains Vijaykumar Gogi of the Karnataka Forest Department. 'The reservoir is a large impediment when it's full,' he adds, 'and we still don't know how the dam and reservoir have impacted migration on a large scale.'

Mobility-wise, the reservoir may negatively impact elephants, but in another respect it helps them. During bone-dry summer months, elephants from throughout Nagarhole and Bandipur travel to the backwaters of the reservoir in search of water and fresh grass. According to *The Hindu* newspaper, the backwaters host the world's largest concentration of wild Asian elephants during the dry season.[8] The magnificent congregation attracts tourists and happens almost every year. Because it brings together a variety of herds, the seasonal congregation gives elephants the chance to find mates. In turn, the landscape-wide population (aka metapopulation) has a better chance of maintaining long-term genetic viability.

Unfortunately, as fate would have it, obstacles continue to crop up in the Nagarhole/Bandipur region. State highways, for instance, now run through both national parks. Roads negatively impact wildlife in a number of ways: construction destroys habitat, and road traffic can alter animal behaviour and mobility. Not only that, but accidents harm and kill animals, as we saw with California's mountain lions. Given India's rapid growth, it's no surprise that road networks have rapidly expanded in recent decades, as have vehicle numbers (there are three times more cars in India today than there were in 1990). 'Human–wildlife conflict due to vehicular movement is an increasing concern in India,' wrote scientists in a 2010 paper, adding, 'the need for reconciling economic growth and wildlife protection on India's highways has perhaps never been greater.'[9]

Given India's size and growing population, authorities cannot always divert roads around protected areas, emphasises Ranjitsinh. 'We have to be practical and instead ask ourselves how we can reduce negative impacts on wildlife.' One method – closing roads between dusk and dawn – has been successfully deployed in Nagarhole and Bandipur. By banning vehicular traffic between 6 p.m. and 6 a.m., wild animals are given a break from human disturbance. Studies show that elephants and tigers need the break whereas animals like boar and deer are less bothered by the presence of vehicles. Research in Malaysia revealed a strong correlation between road crossings and time of day: 81 per cent of elephant crossings occurred between 7 p.m. and 7 a.m.[10] Scientist Sanjay Gubbi of India's Nature Conservation Foundation came to a similar conclusion after evaluating crossings in Karnataka. He even documented elephants walking across empty roads with their tails stiffly raised in the air, a clear indicator of stress, suggesting that even traffic-free crossings can disturb sensitive animals.

Night closures got me thinking about corridors and their role in rewilding. Corridors have traditionally been defined as linear stretches of land connecting core areas, but if we prioritise function over form, anything facilitating movement between point A and point B can be considered a corridor. Animals already use the cover of darkness to navigate problematic patches, whether agricultural fields or cities, so why not give them back the night?

As we'll soon see, several Indian conservationists are advocating for what they call 'dark corridors'. Perhaps it's a bridge too far, but corridors of this kind achieve a form of rewilding. Rewilding, at its core, is about restoring nature and natural processes. For the great majority of human history, until the advent of electricity, people were most active during certain periods of the day. We are a diurnal species, meaning we prefer to be awake during the day and asleep at night.

Elephants are also diurnal, but wild herds living in human-dominated landscapes often *become* nocturnal to avoid confrontations with people. Tigers, meanwhile, are 'mostly nocturnal', say experts, and the same goes for leopards and lions. Dangerous encounters between these species and people frequently occur at night, when animals are more active and people are less likely to see, and thereby avoid, danger. Returning the night to wildlife makes sense, therefore, but in parts of India, the night never stopped belonging to animals. 'Rural India, even today, shuts down after dark,' emphasises carnivore expert Vidya Athreya. 'Most people are asleep, and the landscape is dark. It's almost like wild lands.'

Vidya doesn't ascribe to traditional corridor framings. Animals, not people, use them, so shouldn't animals define what wildlife corridors are (and aren't) and dictate their location? Having collared and tracked leopards and tigers in human-dominated landscapes, Vidya knows that big cats often move through villages and agricultural lands at night. Elephants do the same.

Even in areas with traditional linear corridors, elephants sometimes opt to pass through private property. Of course, some do so to raid crops. In other cases, motivations are less clear. Varun Goswami – who co-founded NGO Conservation Initiatives with Divya – tells me that elephants are so intelligent, they conduct risk assessments before selecting routes. 'They generally move through forested areas because risks are lower there, but they might choose a tea estate over a forested corridor if the estate offers a more convenient path.'

Tea estates have an important role to play, argues Varun. Unlike rice paddy fields or sugar cane plantations, tea estates don't attract hungry elephants. Elephants dislike the taste of tea plants and don't, therefore, crop-raid while moving through plantations, although this doesn't mean their sojourns are problem-free. Crops are often damaged, and people can be harmed if not killed. Rather than trying to keep elephants out of tea plantations and neighbouring villages, Varun and his colleagues are focused on minimising conflict and encouraging coexistence.

Varun borrowed from Sri Lanka's fence-in philosophy when placing solar-powered lights around settlements located within tea estates in the Kaziranga-Karbi Anglong landscape of Assam in the north-east of India. Lights add a layer of safety. Elephants aren't afraid of lights, explains Varun, but they form associations between lights and people, and try to avoid human spaces when possible, making them more likely to pass through an estate under the cover of darkness than wander through areas where people live. Needless to say, locals benefit from solar-powered lighting, as well, since it helps them see potential dangers.

'Once people realised how well lights were working, they asked if we could put them all over the estate,' Varun shared with me. He had to explain that the system only works if elephants are able to access 'dark corridors', a term he and his team coined. Like carnivore expert Vidya, he favours an expanded view of corridors, at least when it comes to India's wildlife. In his opinion, conservationists need to think outside the box when securing connectivity. Government-managed protected areas and traditional corridors remain important, but creative approaches, like dark corridors, can complement conventional ones.

Tea estates in the Kaziranga–Karbi Anglong landscape have been conflict-free since lights were installed in 2017, and locals have demonstrated an astonishing degree of openness towards elephants passing through the area. A survey conducted between 2015 and 2017 found that 96 per cent of 2,500 tea estate workers were happy for elephants to walk through estates as long as human safety concerns continue to be addressed.

In addition to lights, conservationists are using text messages to notify people about approaching elephants. First developed in Valparai, Tamil Nadu, the system has since been rolled out in Hassan, a conflict-prone district in Karnataka. For years, forestry officials attempted to keep people safe by translocating crop-raiding elephants, but like in Sri Lanka, many animals returned. In 2014, for instance, 22 elephants were moved elsewhere, but they all came home in short order.

Hassan's warning system covers 70,000 hectares (172,974 acres) and supports 50,000 people living in proximity to elephants. Text message alerts are joined by automated voice calls and flashing lights at public locations. By encouraging people to head indoors when elephants turn up, alert systems are a bit like dark, or night-time, corridors. Both methods attempt to create a healthy separation between elephants and people, but instead of corralling elephants – the approach used by khedda and translocations – alerts and dark corridors target *human* movement. Doing so is wise given that people are far more likely than elephants to abide by restrictions on mobility, as Hassan's warning system reveals. Before it launched, elephants killed approximately five people per year. Since it went live in late 2017, the annual average dropped to two, and in some of these incidents, victims received but ignored alerts. In 2021, a rash of incidents occurred in areas outside the alert zone after regional coffee estates erected fences, driving elephants elsewhere. 'These events warrant expanding our system,' underscores scientist Ananda Kumar.

Conflict mitigation is only one piece of the puzzle, emphasise conservationists. Habitat must be restored – or at least preserved. Hassan's story bears this out. Scientists monitoring the region's elephants documented a worrisome shift between 2015/16 and 2016/17: during the second year, elephant movement was more restricted, causing an uptick in conflict. Scientists later discovered why. Elephants lost a food source when trees were cut down in an abandoned coffee estate in the middle of the study area. Not only that, but solar fences erected around the former estate restricted wildlife from passing through. Had the land been left to its own devices or, better yet, restored, conflict could have been averted. The incident highlights a point made by Ranjitsinh, 'Rewilding in India is difficult given demand for land, so we have to save what we have, both within the protected area system and outside of it.'

Keeping unprotected land intact is particularly important when you consider the frequency with which elephants move beyond protected areas. A study in Karnataka found that elephants regularly utilised unprotected areas containing dense human populations. In fact, 60 per cent of the places elephants used were outside protected area boundaries, causing the study authors to question India's conservation framework. Because it 'legally protects the species wherever it occurs, but protects only some of its habitats', India's framework 'seriously falters in situations where elephants reside in and/or seasonally use areas outside' protected areas.

While Varun and Vidya explore creative corridors, organisations like the Nature Conservation Foundation and Wildlife Trust of India (WTI) expand conventional cores and corridors. Before returning to traditional corridors, let's spend a minute on cores. As I mentioned earlier, southern India's Nilgiri Biosphere Reserve, while critically important, isn't a protected area per se since only 22 per cent of its 552,000 hectares (1.4 million acres) enjoys strict legal protection. However, that percentage has risen during recent years and continues to climb, thanks to the work of conservationists like Sanjay Gubbi, whom we met earlier when discussing road impacts on elephants.

A native of Karnataka, Sanjay believes that land tenure is everything in India. If land isn't secured for wildlife now, 'in the future it will be gone'. The biologist has spent decades fighting to increase the state's protected area network. His approach is simple and cost-effective: he works with the government to convert what are known as 'reserve forests', which have nominal protection, into sanctuaries and national parks, both of which enjoy strict legal protection under India's Wildlife Protection Act. Because the government owns reserve forests, giving them a higher level of protection is mainly an administrative matter. Doing so is a no-brainer, says Sanjay, but the process can be challenging.

From Sanjay's perspective, only 10 per cent of conservation boils down to hard science. The rest hinges on communication and relationships. 'Getting the proposal right is the first step,' emphasises the biologist, 'then you have to get it to the right people and push it through the entire hierarchy.' Being a local with longstanding relationships is key, in his opinion. Although the model he developed is now utilised across India, he would never work in other states. 'India is so vast and so diverse, and conservation is a very social and political field.'

During an August 2021 Zoom interview, Sanjay used a series of maps to walk me through the changing status of wilderness areas in the Nilgiri Biosphere Reserve. I could clearly see how things looked when his work began 20 years ago: although the reserve contained a number of fully protected areas, they were separated by large swathes of unprotected land, much of it intact and forested. Sanjay then showed me the map as it stands today. A higher percentage of it is covered in dark green, the colour denoting protected areas, and if you zoom into Nagarhole/Bandipur you see that the park system has gained protected land in recent years (22,500 hectares (55,599 acres), to be exact). India-wide and over the course of two decades, Sanjay helped convert 300,000 hectares (741,316 acres) of forest reserves into protected-area land. Twenty years ago, Karnataka's protected-area network comprised 3.8 per cent of the state; that figure has since grown to 5.2 per cent. The expansion is the largest in India since the 1970s, says Sanjay, but his work continues. He is hoping to build a 1.3 million-hectare (3.3 million-acre) network of protected areas that form a single landscape in the Western Ghats.

In the meantime, the WTI fills in gaps by buying up smaller parcels of privately owned land and converting them into traditional corridors. Years ago, WTI and the Asian Nature Conservation Foundation teamed up with state governments and elephant researchers to identify existing corridors, publishing their findings in a 2005 report. The picture was bleak. Although 88 corridors existed, few provided safe passage for elephants and other animals. Human settlements were found within 78 per cent of the country's corridors; 20 had rail lines cutting through them, and highways passed through 66 per cent. Only a third were free of agricultural activity.

Notably, elephant corridors aren't currently protected under Indian law, so conservationists couldn't simply petition the government to enforce existing protections. Instead, they had to create new ones. In the years that followed WTI's report, a range of entities – from non-profits to local communities and state governments – began buying up land that fell within elephant corridors. Such corridors, according to WTI, are 'linear landscape element[s]' that 'facilitate movement across habitat patches'.[11] From the organisation's perspective, linear landscapes only qualify as corridors if they connect 'source patches' capable of sustaining stable or growing populations. Function is as important as form and one can't exist without the other. WTI wouldn't, therefore, consider 'darkness' a type of wildlife corridor. Interestingly, the NGO discourages habitat restoration *within* corridors since doing so might encourage elephants to stick around. Because corridors are often located beside communities, the faster elephants move through, the better.

In some cases, buying land is the first and final step towards ensuring a corridor's protection. Other times, additional actions are required. To get a sense of them, it's worth turning our attention to several corridors in the Nilgiri Biosphere Reserve. One, the Thirunelli-Kudrakote Elephant Corridor, connects Karnataka's Brahmagiri Wildlife Sanctuary with Kerala's Wayanad Wildlife Sanctuary, both of which are located to the west of the Nagarhole/Bandipur park system. WTI calls the 890-hectare (2,200-acre) corridor a 'lifeline' for the region's elephants, and the passageway is also used by tigers and other animals.

Part of it sits on protected land, but when WTI identified the corridor in 2005, portions were occupied by people or set aside for agriculture. Working with the Kerala Forest Department, WTI purchased 10 hectares (25 acres) of corridor land from local communities who were then encouraged to voluntarily relocate to sites within a 5-km (3-mile) radius (voluntary relocation of this kind is common in India, and proponents of it argue that locals benefit by moving to areas with reduced conflict and greater access to infrastructure, but critics point out that 'voluntary' resettlement is often coerced if not forced). In total, 32 families from four villages opted to move and were either paid in cash for their land or given a relocation package that included a new home and agricultural land.

Working with a local NGO, WTI subsequently surveyed individuals who opted to relocate to see whether their lives – and attitudes towards conservation – had changed. Annual income had grown, as had access to potable water. Now residing closer to schools, more children were pursuing an education, so literacy rates had gone up. Human–wildlife conflict, meanwhile, had all but disappeared. Families continued extracting non-timber products, like mushrooms and tubers, from the forest, but dependency on such products decreased following relocation. Most interesting, in my opinion, were attitudes towards conservation. After relocating, 12.5 per cent of the community felt 'very positive towards conservation' and forest preservation. The great majority (85 per cent) were indifferent towards conservation – while 2 per cent held negative attitudes towards conservation. Unfortunately, these questions were not posed to families before they moved, so there is no way of knowing whether voluntary relocation changed attitudes.

On the eastern side of the Nilgiri reserve lies the Sigur Elephant Corridor. Comprised of four individual corridors, the wider pathway protects critical east to west elephant movement along the Sigur Plateau. Declared an Environmentally Sensitive Area, the plateau links numerous national parks and wildlife sanctuaries, among them Nagarhole and Bandipur. Conservationists have been fighting for the Sigur Elephant Corridor's creation for years, but human rights groups have pushed back on the basis that it would displace some 12,000 people. The battle ultimately made its way to India's Supreme Court, which in October 2020 upheld a lower court decision authorising removal of human settlements from the corridor.

In effect, elephants won the right to unfettered use of the land. 'It is probably the only case in Asia where the highest court put the needs of elephants first,' says WTI's Avinash Krishnan, making it a landmark – yet highly controversial – decision. Whether resort owners and families had legal title to the land they occupied is under dispute, but there is no denying that people will suffer economic, and perhaps cultural, losses as a result of the decision. It highlights that compromise is not always available when the needs of animals and people conflict.

Death, economic loss, eviction and resettlement: these are some of the costs borne by people who live alongside elephants. Elephants, of course, pay steep prices of their own, but unlike people, they have no way to petition the government for compensation or support. Managing human–elephant conflict – or 'interactions', a term preferred by some of today's conservationists – is no easy task, but one thing seems clear. Asking people to change locations is easier than encouraging elephants to chart new paths.* China's mammoth wanderers proved this time and again as they ignored plush relocation packages while moving unpredictably. In the Nilgiris, at least, scientists have decades of data pointing to well-trodden elephant routes between established protected areas. Knowing what to protect is half the battle.

September 2021

Just before I finished writing this chapter, I learned that China's elephants had finally come home. Eighteen months had passed since the herd left its native habitat and began the circular journey that captivated the world. Despite traversing hundreds of kilometres, and making pit stops in villages and cities, the herd managed to complete its journey without harming any people – thanks, in large part, to Chinese authorities that monitored the elephants around the clock. Confident that the herd's trip had come to an end – at least for now – provincial authorities wound down their intensive surveillance programme. They called the herd's homecoming 'very meaningful' and were quick to point out that the experience 'shows China's resolve in building an environmentally friendly country'.[12]

Motivations aside, the Chinese government successfully navigated a perilous situation in a way that kept elephants

---

* Asking humans to move may be (comparatively) easier than forcing elephants to adapt to new terrain, but neither approach is straightforward, particularly in a place like India. As Sanjay Gubbi pointed out in a paper on tigers, 'Meeting the competing demands of development and the preservation of wildlife habitats is the biggest challenge faced by governments, managers and conservationists in a developing country such as India, where significant changes in land use are occurring.'

and people safe. We may never know *why* the herd travelled where it did, but the many episodes detailed in this chapter highlight that large, wide-ranging animals, like elephants, have been hemmed in by human activity. Ensuring these animals have ample habitat and resources is critical to their survival – and ours. Which is where cores and corridors – conventional and creative – come in. It didn't occur to me until later, but by evacuating villages and corralling elephants (or at least attempting to), Chinese authorities were creating a corridor of sorts, in real time, as the herd headed north. Needless to say, few countries have adequate resources to build impromptu makeshift corridors, but the concept, at least, is intriguing.

I wondered what takeaways, if any, emerged from China's story. Would elephant experts recommend crafting corridors based on the herd's movements? I put this question to several biologists including Pruthu Fernando. The Sri Lankan does not believe that China's herd was moving in a natural or predictable fashion. He might reconsider his stance 'if the elephants keep doing the same trek again and again', but the one-off, haphazard journey does not, in his opinion, justify the creation of new corridors. Another scientist, who would only speak to me off the record because she does not work in China, underscores that corridors are only viable when they link islands of suitable habitat. Based on her understanding of events, the elephants were not travelling between habitat cores or patches, so designing a corridor based on where they went makes no sense.

CHAPTER TEN

Community Conservation

Sera Community Conservancy, northern Kenya, May 2015

Four rangers clad in army fatigues stood atop a massive metal crate containing precious cargo. Twelve metres away, National Geographic wildlife photographer Ami Vitale readied her camera for a series of action shots. Three, two, one! The rangers heaved the crate door open. Moments later, a black rhino leapt forward and crashed into the open air. Mini-tornadoes of red dirt swirled around its feet as the 700kg (1,543lb) animal sprinted towards a cluster of trees.

It was a momentous occasion; one that took years to achieve. When the first of 13 rhinos released into the community-owned Sera Rhino Sanctuary ran free, local Samburu communities erupted in song and dance as they celebrated the return of black rhinos to their homeland after a 25-year absence. Many community members had never seen rhinos before. According to Ami, 'Some thought they would have soft skin (like a cow's) and horns that were as flexible as an elephant's trunk.' Elders, on the other hand, clearly remembered encountering rhinos decades earlier. Mzee Master Leodip, now in his 70s, regularly came upon them during his youth. He crossed paths with poachers, too, when they approached offering him sweets in exchange for information about rhino whereabouts. Mzee stoically recounts the animal's decline: 'The rhinos became fewer, and then they vanished, and we never saw them again until 2015.'[1]

Conservationists have cause to celebrate whenever they help reinstate part of an endangered species' historic range, but the 2015 reintroduction of black rhinos to the Sera Rhino Sanctuary made headlines for a different reason, one relating to stewardship. For the first time in East Africa, a local community was given the opportunity to look after the critically endangered black rhino on community land. The rhino sanctuary, you see, is part of the larger community-owned Sera Community Conservancy. A *Mongabay* article called the handover 'unprecedented'[2], while author Peter Martell, who recently published a book on Kenya's northern rangelands, told me it required a massive leap of faith. From Ami's perspective, the media frequently covers endangered species conservation whereas 'little has been said about the indigenous communities on the frontlines of the poaching wars'.[3]

When I think about wilderness areas, I instinctively picture national parks and reserves. Like Doug and Kris Tompkins of Chapter Seven, part of me ascribes to the philosophy that national protected areas are 'the most durable way to protect wildlife habitat and help people reconnect with nature'.[4] At the same time, I believe that local people make nature's best guardians. While these two beliefs don't have to be mutually exclusive, they often are in practice. Again, I think of Yellowstone, whose creation saw indigenous people pushed off their land. America's first national park inspired protected areas all over the world while establishing the notion that people and wilderness are incompatible – or at least separate. Wilderness, the model holds, is untouched, unspoiled. Never mind that the 'wilderness' Yellowstone sought to preserve was already modified by indigenous people.

In Kenya, as in many parts of Africa, land was communally owned for centuries, and in arid environments like the northern rangelands (where Sera is located), indigenous communities practise pastoralism. Poor soils and frequent droughts make crop cultivation difficult, if not impossible, so people keep livestock instead. They live semi-nomadic lives while herding livestock towards green(er) pastures. Livestock, in turn, provides income, status and cultural identity. As you can imagine, private, fixed land ownership makes little sense for groups of people living nomadic or semi-nomadic lifestyles.

Globally, upwards of 500 million people practise pastoralism – and all of northern Kenya's tribal communities do so. For these groups, geography has historically revolved around livestock grazing. Within Kenya, pastoralists are the minority – 75 per cent of Kenyans live in areas conducive to productive agriculture – but 80 per cent of Kenya's landmass is arid or semi-arid, 'making pastoralism arguably the most expansive land use in Kenya'.[5] Despite this, the British colonial government and post-independence Kenyan government marginalised pastoral lifestyles and community land ownership practices. Pastoralism was minimised, at least in part, because uncultivated land was considered worthless.

Several mechanisms for communal land ownership were put into place following Kenya's independence from Great Britain (in 1963), but they were problematic, say experts. It wasn't until 2010 – when the Kenyan government enacted a new constitution – that community land rights were officially recognised. Another key milestone occurred soon after when the Kenya Wildlife Service (KWS) acknowledged the need for community involvement in a black rhino conservation strategy report published in 2012. 'Engaging the communities through dialogue, support and improved awareness on the plight of the rhino will be essential to increase community participation in conservation and protection of rhinos,' wrote the report's authors.[6] Community participation was imperative, in part because existing protected areas were running out of space for endangered wildlife, including rhinos, as populations rebounded following years of dedicated conservation. Then, in 2013, the country passed legislation formally recognising community conservancies.

By the time the 2012 KWS rhino strategy report called for community involvement, the Sera Community Conservancy was on its way towards opening the first community-run black rhino sanctuary in East Africa, but question marks remained. Did Sera contain ample black rhino habitat? Was it secure enough to stave off poachers? And would community members, many of whom feared black rhinos, welcome the species back with open arms?

Sera Community Conservancy's 340,000 hectares (840,158 acres) used to be a hotbed of tribal conflict. Sera borders two indigenous communities – Borana and Samburu – that were often at war over, among other things, livestock grazing rights. To avoid conflict, herders avoided the land on which Sera lies except when absolutely necessary. Times of drought brought Borana and Samburu together, causing clashes that frequently centered on livestock raids and sometimes turned violent. Mzee remembers spending mornings sharpening his machetes while 'waiting for a signal to track down stolen livestock.'[7] Some of the men sent on cattle retrieval missions never came back. For Sera, however, conflict had an upside. Because it was off limits most of the year, the land remained relatively undisturbed by human activity. 'Sera stayed wild,' says Kieran Avery of the Northern Rangelands Trust (NRT). 'But it was also the wild west.'

Sera was formed (in 2001) to de-escalate ethnic tension, says Sera Conservancy manager Reuben Lendiro.

Peace-making was goal number one, but Reuben and others in his Samburu community knew that conservancies had the potential to create new financial opportunities. In turn, local livelihoods would improve. Wildlife tourism, in particular, can transform communities, as we've seen in numerous chapters. However, tourism revenues often 'leak out' of the areas where tourism takes place, instead landing in the pockets of foreign companies. Community conservancies are a way to balance the scales while incentivising environmental stewardship.

For Kenya's wildlife, community conservation is essential. Although the country contains numerous national parks and reserves, some world-famous, two-thirds of Kenya's wildlife is found *outside* government-managed protected areas on land containing people. Communities have no choice but to coexist with animals living beyond the perimeters of national parks and reserves. According to the Kenya Wildlife Conservancies Association (KWCA), 'People living alongside wildlife hold the power to determine whether wildlife will flourish or perish.' They bear the costs that come with living beside wild animals but simultaneously stand to benefit from conserving wildlife – *if* they are empowered to do so.

Community conservancies have become a leading empowerment tool and are on the rise. Kenyan law defines them as areas of land set aside by communities for wildlife conservation. Here it's worth noting that other kinds of conservancies exist, as well: private individuals, groups of people and corporations can all create conservancies. KWCA underscores that community conservancies offer the best legal structure for enabling local communities to 'manage and benefit from natural resources'.[8] Several studies highlight their positive impacts, which range from enhanced security to increased household income.

To evaluate the effects of community conservation, one study compared three northern Kenyan community conservancies, including Sera, against non-conserved sites with similar socio-economic and environmental conditions. The study's authors concluded that conservancies in northern Kenya are positively impacting communities and the environment. 'Conservation has enhanced livelihoods by facilitating community access to public services and infrastructure,' they wrote, adding, 'these socio-economic changes have occurred in the context of significant improvements to habitat condition driven by sustainable grazing management.'

Community conservancies can generate income through multiple revenue streams – of them, tourism is the most valuable. In the words of Kenyan writer Paul Udoto, 'Wildlife tourism is the proverbial goose that lays the golden egg in the Kenyan economy – it's the lifeline.' It's been that way for over a century, first attracting trophy hunters and, later, safari-goers (Kenya banned trophy hunting in 1977). In 2019, before the pandemic turned the industry upside down, tourism brought $1.6 billion into the country.

However, just because a community establishes a conservancy doesn't mean it gets a slice of the pie. To attract tourists, conservancies need infrastructure and security on top of wildlife. Since Sera was established on no man's land, many wildlife populations were intact when the conservancy was born in 2001. Rhinos, however, had vanished. Poachers killed the last one in the early 1990s.

At the time, black rhinos were crashing towards extinction. Across the board, Kenya's wildlife was devastated by decades of hunting and poaching (between 1977 and 2016, Kenya lost an estimated 68 per cent of its wildlife).[9] Black rhinos were nearly wiped out. As we learned in Chapter Six, the black rhino, or *Diceros bicornis*, is the most endangered of Africa's two rhino species. Its numbers fell by 96 per cent between 1970 and 1992, mainly as a result of hunting and poaching. Kenya's black rhinos all but disappeared in the span of two decades, falling from 20,000 in the 1960s to as few as 300 in 1990.

Kenyan wildlife officials were aware of the herbivore's decline and recognised that if action wasn't taken, the few remaining populations, which were highly fragmented, would die out. So, in 1984, the government launched a new initiative, one that entailed moving rhinos to secure sanctuaries where they could live in peace and breed. KWS already had experience moving rhinos between regions and, by this time, several pioneering private ranches had begun offering safe harbour to black rhinos.

One, the Solio Ranch Game Reserve of southern Laikipia, started stocking its fenced black rhino sanctuary before the term 'rhino sanctuary' was even coined. During the late 1960s and early 1970s, Solio rounded up 23 rhinos, relocating stragglers from various parts of Kenya to its relatively small fenced sanctuary. Solio's population grew rapidly. It boasted 'an exceptionally high annual birth rate of 15 per cent from 1980–86,' wrote Rob Brett in 1988. The conservationist now works for Fauna & Flora International, but at the time he was an independent researcher in Laikipia. The following year, KWS hired him as its rhino coordinator, a position that involved overseeing the stocking of rhino sanctuaries across Kenya.

Soon after Solio created its sanctuary, another ranch followed suit. Lewa Downs, or Lewa as it's known today, is also located in Laikipia. Lewa entered the realm of rhino conservation when long-time wildlife warden Peter Jenkins began advocating for sanctuaries on private land. Concurrently, English conservationist Anna Merz

approached Lewa owner David Craig about establishing a
rhino sanctuary on his family's 25,091-hectare (62,000-
acre) land. Merz was compelled to act after witnessing the
black rhino's destruction (when she visited Kenya in 1968,
she saw black rhinos 'everywhere', but eight years later,
their vibrant bodies were replaced by corpses in a massacre
she called 'unbelievable' and 'horrifying'). Merz was so
determined to help, she used a $750,000 family inheritance
to finance the creation of the Ngare Sergoi Rhino Sanctuary
at Lewa. The 2,023-hectare (5,000-acre) sanctuary broke
ground in 1983.

Within a matter of months, Merz and the Craigs
welcomed their first rhino, a male from Nairobi National
Park. Additional translocations followed as the Craig family
collaborated with the Kenyan government, expert pilots,
trackers and vets to transfer northern Kenya's remaining
black rhinos to the sanctuary. New arrivals were kept in
temporary holding pens as they acclimated to their new
environment. According to her *New York Times* obituary,
Merz used to sit beside the pens reading Shakespeare aloud,
a tactic she claimed soothed the rhinos' nerves.

Merz's sanctuary was a remarkable success. So much so
that the Craig family decided to convert the entire ranch
into a private conservancy in 1995: the Lewa Wildlife
Conservancy was born. Black rhino numbers steadily grew
– from 15 at its founding to 126 today – but Lewa doesn't
just shelter black rhinos. Together with a neighbouring
conservancy, Borana, Lewa's rolling hills are home to 112
white rhinos, more than 400 elephants, lions, cheetahs and
the endangered Grevy's zebra (the Lewa/Borana landscape
contains 11 per cent of the global population of the thin-
striped, large-eared equine).

As rhino numbers climbed, Lewa began mulling over
ways to support conservation beyond its borders. Rhino

translocation was one method. Lewa has moved 54 black rhinos to other sites in Kenya, among them Ol Pejeta – the northern white rhino sanctuary where this book began – and, later, Sera. Lewa also recognised that for conservation to work in the long term, 'it needed to involve and benefit the pastoralist communities in the surrounding areas'.[10] To kick-start community collaboration, Lewa owner Ian Craig (David's son) invited elders from a nearby Maasai community, Il Ngwesi, to visit the conservancy in 1995. Craig hoped the Il Lakipiak Maasai – which means 'people of wildlife' – would see that conservation, ecotourism and livestock ranching can go hand in hand. Il Ngwesi's elders were sold. The following year, they dedicated 8,675 hectares (21,436 acres) of grazing land to conservation while building a luxury eco-lodge. It is, according to the conservancy, the only upmarket safari lodge owned *and* managed by a Kenyan indigenous community.

A second northern Kenyan community conservancy, Namunyak, was born soon after. Sera and others quickly followed. Community conservation was gaining ground in the north of the country and beyond, but communities needed support to become operational. Governance and land rights posed challenges, as did fundraising. Wildlife conservation, too, was new for many groups. Helping communities get the model right was in everyone's interest. Besides, the Kenyan government (via KWS) manages the country's natural resources, including wildlife, on the people's behalf. Black rhinos and other critically endangered species wouldn't return to conservancies like Sera without the government's blessing – and involvement.

To scale up community conservation, Lewa created a new entity – the Northern Rangelands Trust (NRT) – in 2004. NRT, which is independent of Lewa but headquartered on the former cattle ranch, serves as an umbrella organisation for community conservancies in the northern rangelands. Both NRT and the conservancies it helped establish have rapidly grown since its founding 18 years ago. The trust currently employs 225 people (including Kieran, whom we heard from earlier) and supports 39 member conservancies (some but not all of which are managed by communities). Northern rangelands members represent around 40 per cent of Kenya's conservancies. Another large cluster is located in the Maasai Mara.

Seeing the benefits community conservancies offered, some communities immediately warmed to the idea of launching their own. Others needed convincing.

Isa Gedi was born in the village of Hara in north-eastern Kenya's Ijara District in 1986. As a boy, Isa spent many of his weekends tending to his father's cattle. On scorching walks through arid bush, he spotted topi, warthog and zebras, as well as large herds of hirola antelope. He would later learn that the long-faced antelope was rapidly disappearing from the planet, but at the time he considered the region the Maasai Mara of the north.

Isa loved watching wildlife. During year six of primary school, his teacher asked students to share their plans for the future. 'I want to look after cows and watch wildlife graze,' proclaimed Isa. His teacher laughed before responding, 'If you want to make something of yourself, you have to leave this village.' Deep in his gut, Isa knew that jobs guarding wildlife existed. He also knew that he wanted to work with communities, but he needed time – and experience – to understand how to make these goals converge.

Like his teacher, Isa's parents pushed him to pursue a stable, conventional career, but he turned down a government-funded university degree in teaching in favour of a community resource-management programme, even though it meant securing funding to cover his tuition. After graduating in 2012, he returned to his village while preparing to move to Somalia to work for the Red Cross. While there, he heard members of his community discussing conservation. They were heatedly debating whether to create a hirola antelope sanctuary inside the Ishaqbini Community Conservancy they established years earlier – in 2007.

Safeguarding the species, which is only found near the border between Kenya and Somalia, was urgent because hirola numbers had fallen from around 16,000 in 1979 to fewer than 500. This, despite the fact that the governments of Kenya and Somalia both instituted hirola conservation programmes in the 1970s. Declines were predominantly caused by drought, habitat degradation and a rinderpest epidemic in the 1980s (the viral disease can pass from cattle to wildlife). Interestingly, one of the reasons the antelope's preferred habitat – grasslands – diminished owed to decreasing elephant numbers. Hirola conservation needed a new approach. Many members of Isa's community were interested in pursuing conservation, but others were unwilling to give up their grazing land.

The more Isa listened to the debate, the more certain he was that conservation was the only path forward, and the fledgling conservationist quickly came to another conclusion: he had a role to play in jump-starting community conservation. He was due to begin his Red Cross job in Somalia two weeks later but felt compelled to take another path – one that could help save the hirola while changing lives at home. From Isa's perspective, locals were wavering on establishing a hirola sanctuary for one of two reasons: they fundamentally misunderstood the community conservation model, or they questioned the motives of established conservationists, like those at NRT.

Because he was a respected member of the community and someone who possessed conservation knowledge, Isa felt that he could help persuade his peers, but first he needed to gather additional information about community sanctuaries so he could present an airtight case. After giving up his job with the Red Cross, Isa began travelling around Kenya to meet with different conservancies. Although he self-financed his research, he communicated his aims to NRT and later went to work for the organisation. After visiting six or seven conservancies, Isa drafted a 15-page report laying out the community sanctuary model. Back home, he presented his findings at five public meetings and sat down with local chiefs to encourage them to move forward. Together with data gathered by Ishaqbini rangers during the preceding five years, Isa's presentation helped the community reach a consensus.

The Ishaqbini Hirola Sanctuary was formally established in 2012 as a predator-free, solar-powered electric fence came up. Forty-eight hirola from the surrounding area were rounded up and brought into the fenced sanctuary. A year later, 48 had jumped to around 120, and since then the herd has annually grown by an average of 13 per cent. Before long, the sanctuary reached carrying capacity, so staff began transferring animals (males, mostly) to neighbouring rangelands. Unfortunately, the wild antelope living outside the safe harbour of the sanctuary continue to decline, so conservationists are thinking about increasing its size. Outsiders have proposed translocating animals from the sanctuary to other parts of Kenya, but community members aren't prepared to part with any of their antelope. 'Locals believe the hirola is a blessing from God,' explains conservancy manager Ahmed Maalim. 'The hirola's presence, for them, signifies prosperity.'

Several hundred kilometres away, Reuben and his colleagues at Sera were busy building a fence of their own. When completed, it would enclose the Sera Rhino Sanctuary, which is where our story began. As in Akagera (Chapter Six), the wild animals already living in the area were fenced in. Constructing the fence was an arduous task that took two and a half years, but the work that preceded it was far more challenging. Habitat and security assessments had to be conducted before the Kenya Wildlife Service would even think about sending rhinos to the area, and funding – for the fence, security and translocation – would have to be secured.

First, community members needed to decide whether they wanted to bring rhinos back to their land. Elders remembered what life was like when people and rhinos coexisted. They knew people who were injured or killed by the mega-herbivore and worried that the rhino's return would compromise their safety. A more common concern centered on grazing: by fencing off a portion of the conservancy, locals wouldn't be able to take their livestock wherever they wanted – a hindrance that would be acutely felt during dry seasons when grasses and water are scarce. Conservancy manager Reuben Lendiro countered that black rhinos are browsers, not grazers, so would not, therefore, consume the grasses that sustain cattle. He also pointed out that rhinos drink less water than cattle. In any case, livestock and cattle would rarely mix since the land set aside for the wildlife sanctuary would be off limits to locals and their livestock except during times of severe drought. Even then, livestock could only be taken to specific areas that do not overlap with rhino territory.

Pros and cons were weighed, and debates waged. Rhino rewilding would not move forward without community approval. Ultimately, as the start of the chapter reveals, Sera's owners came to see that the benefits of reintroducing rhinos outweighed the costs. As someone who is endlessly fascinated by creatures great and small, I would love to tell you that communities agreed to bring rhinos back because they revere the animal and wanted to make their ecosystem whole again. In reality, the decision appears to have boiled down to practicalities. Because poaching remains a threat, areas containing rhinos require considerable security in the form of trained rangers, anti-poaching units, electric fencing, etc. The people living in Sera's proximity knew that enhanced security would not merely benefit rhinos: it would help keep them and their livestock safe from tribal and other disputes. At the same time, communities understood that black rhinos would bring tourists – and revenues – to Sera.

Let's go back to May 2015 when photographer Ami Vitale documented the black rhino's return to Sera. In total, 13 black rhinos were released into the conservancy's fenced sanctuary, which spans 10,700 hectares (26,440 acres), but conservationists now refer to a founding population of 10 animals. That's because three of the rhinos sent from Lewa and Lake Nakuru National Park to Sera perished as a result of post-translocation complications. Moving animals is always risky, particularly when they have to be sedated, and rhino deaths have sadly occurred as a result of numerous translocations.

During a particularly devastating 2018 incident, 11 black rhinos moved to Tsavo East National Park died during or after translocation. As NGO Save the Rhino pointed out, the death toll exceeded the number of rhinos killed by poachers across Kenya during the two preceding years. An enquiry into the incident concluded that human negligence played a part in the animals' deaths; translocation-induced stress was another factor. There is no excusing a loss of this magnitude, particularly when human error is involved, but endangered species conservationists seem willing to tolerate *some* risk – even loss.

Translocation guidelines published by the IUCN Conservation Translocation Specialist Group state that 'the loss through death or dispersal of translocated individuals at some sites or in particular years should be expected'.[11] Every year, the group publishes a book detailing recent species reintroductions – and the 2018 edition analysed Sera's black rhino reintroduction, deeming the project a success in spite of rhino deaths. Five reasons were given, four of which relate to community conservation. 'Convincing the government' to allow 'a local community to host and manage highly protected wildlife (black rhinos) outside national parks and private sanctuaries' was considered 'a big risk', wrote the authors.

Sera's rhino population may have started out smaller than expected, but it has grown by 20 per cent each year and now stands at 19. Within a year of settling in, Sera's rhinos began mating, and the first calf conceived at Sera was born in 2017. Remarkably, no poaching incidents have occurred since rhinos arrived. Biologist Antony Wandera, who has worked for KWS and NRT, says the threat is ever-present. 'Every other day, we deal with new, emerging threats,' he told me during an August 2020 interview. He credits law enforcement and a tight regional security network for keeping poaching at bay, but community policing is invaluable. 'Because of land ownership, locals are at the forefront of protecting animals,' Antony explained. 'No one can get through.'

These days, the communities that own and manage Sera have even more reason to maintain the integrity of their land. In May 2020, another critically endangered species made its way to Sera when Lewa sent 25 Grevy's zebra to Sera's Rhino Sanctuary. It was the first ever translocation of the species to a community protected area – another indicator that Sera and conservancies like it are fundamental to long-term wildlife conservation. Known for their thin stripes and large body size, Grevy's zebras are found in Ethiopia and Kenya, but have gone extinct in Somalia and Sudan. Around 15,000 survived in the 1970s; today, just over 3,000 remain. Northern Kenya is their stronghold. Sera already had 17 Grevy's on-site when the May 2020 translocation occurred. Without an injection of new blood, the small population was unlikely to thrive, and Lewa had Grevy's to spare. Wildlife veterinarian Mathew Mutinda called the reintroduction incredible, adding that Sera is 'one of the most genuinely successful conservancies' he's encountered, 'an example of how you can build something from nothing'.

At the nearby Namunyak Community Conservancy, another coalition of Samburu communities came together to build something remarkable. They constructed the Reteti Elephant Sanctuary, Africa's first community-owned elephant orphanage, which opened in August 2016. There, dedicated keepers, a quarter of whom are women, look after elephant calves rescued from the northern rangelands and prepare them for life in the wild. When orphans are ready to be rewilded, they go to Sera.

Of the 50 or so orphaned elephant calves that have come to Reteti, Shaba is the most famous. The young female arrived as a 15-month-old orphan in November 2016 after poachers shot her mother. She was transported by helicopter from the Shaba National Reserve and arrived traumatised. For days, keepers tried to earn her trust. 'We spent day and night, talking, singing, offering her seed pods, fresh grass … anything we could think of to win her over,' wrote Shaba's keepers on Reteti's website. 'But she just didn't want anything to do with us. And we didn't blame her. But we needed to convince her that we were here to help.'

Singing to animals is traditional in Samburu culture. Herdsmen sing love songs to their cattle while cows drink from troughs filled with water from the 'singing wells of Namunyak'. 'I will always be with you, I will always be behind you – you are my lovely cows,' sang one herdsman.[12] The tradition has been adapted to elephants. Videos posted to Reteti's Instagram feed show keepers using song to connect with fragile orphans. In one, a male keeper sings a rhythmic chant to a calf as it drinks from a bottle in his hands. Merz, it seems, wasn't the only animal caretaker who believed in art's healing powers.

Because of the pandemic, I haven't travelled to Reteti, but I have witnessed the close bonds linking orphans and human caretakers during visits to Africa's best-known elephant sanctuary: Nairobi's David Sheldrick Elephant and Rhino Orphanage. During my first trip in 2011, I watched a group of elephant calves race towards their keepers for late-afternoon bottles of milk after a full day in Nairobi National Park. Some were tiny, others mid-size, but each of them knew that the men sporting beige hats and green jackets had dinner in hand. After gulping down its evening milk, a tiny orphan ambled past me as it followed a keeper towards a barn-like wooden stall. Once inside, the calf lay down for the night while the keeper grabbed a grey blanket and draped it over the youngster before climbing into a bed built above the floor.

Keepers rotate stalls each night to prevent orphans from becoming too attached to any one person, a lesson the late Daphne Sheldrick, who founded the orphanage, learned the hard way. Tragically, an orphan hand-reared by Daphne died when the conservationist left the orphanage for several days after the two spent months side by side. The absence led to unbearable stress and a broken heart.

Shaba seemed broken-hearted when she landed at Reteti six years ago, but the young calf suddenly perked up one day, and from that point forward she was a different animal. It soon became clear that Shaba had strong maternal instincts, perhaps because she spent 15 months under her mother's care before she was orphaned. As Reteti's leader, its pseudo-matriarch, Shaba taught youngsters invaluable life skills, like how to navigate steep paths, while showing them tenderness. Among other things, she made a point of greeting every orphan upon arrival.

Her keepers were smitten, and because animal caretaking was new to them, many struggled when it came time for Shaba to leave. She was supposed to be part of the first batch of orphans translocated to Sera in May 2019, but Reteti staff worried that removing the maternal figure too soon would disrupt herd dynamics. So, they held her back when transferring three male orphans to Sera.

Before long, however, Shaba began behaving differently. Wildlife veterinarian Stephen Chege told me she became 'notorious during her last few days at Reteti'. 'Shaba was opening doors, breaking down fences and pushing into people.' She was ready to go home – to the wild, that is. It was clear that Shaba needed more space, but her caretakers weren't ready to say goodbye. Stephen reminded them that returning orphans to the wild is the right thing to do. 'I know I will cry, not only today but tomorrow,' conceded keeper Dorothy Lowaktuk in a short documentary film made by Ami Vitale, 'but we can't deny her the right to go back to the wild.'[13] Dorothy said goodbye on 16 November 2019.

Unlike the black rhinos sent to Sera in May 2015, Shaba and her fellow orphans were not part of a species reintroduction. Wild elephants never left Sera. In fact, elephants from Laikipia and Samburu often make their way to Sera during heavy rains because it's a place where they feel safe. A smaller number (around 42) live within the fenced rhino sanctuary. They can't come and go as they please, but if the population outgrows the sanctuary, some animals will have to be translocated or ushered out. Sera and KWS are hoping to add corridors so elephants can freely move between the rhino sanctuary, the rest of Sera and the wider region.

The first group of orphans sent from Reteti to Sera adapted better to their new normal than the two cohorts that followed. They are in better health, spend more time with wild herds and move throughout the entire sanctuary, whereas the seven orphans subsequently translocated tend to stick together and are mainly found in the eastern part of the protected zone. Shaba, unsurprisingly, leads the herd of orphans. She may be a natural leader, but she is young and inexperienced, which may explain why the second and third cohorts use less of the sanctuary. Gender probably plays a part, as well, posits Frank Pope, CEO of Save the Elephants (the Samburu-based research organisation is conducting the first ever study on orphan elephant rewilding). All of the orphans in the first cohort were male, and one of the existing, fully wild herds contained young males of similar ages. 'One way an orphan can get accepted into a wild herd,' explains Frank, 'is by making friends with a wild elephant of its same age.'

As of September 2021, Reteti had translocated a total of 10 orphans to Sera. Eight of them are alive. The other two were killed by lions. 'Rewilding orphaned elephants is a long-term process,' emphasises Stephen, who coordinates all of Reteti's rescues and releases. Lessons are learned along the way. 'The lion attacks caused us to realise that orphans growing up in sanctuaries may be at a disadvantage since they don't grow up learning about threats like lions.' To prepare them for possible encounters, Stephen and his colleagues are brainstorming ways to train orphans to associate predators with danger. They may, for instance, play audio recordings of lions in an approach similar to the one used by Iberá's green-winged macaw keepers (Chapter Seven).

When you think about the dangers facing young animals, you picture predators or poachers, but some threats come from within the family. Lojipu, which means 'to follow', was the first black rhino born in Sera's Rhino Sanctuary. He entered the world in February 2017 and was born to a female from Lewa called Nairenyu, who was already pregnant when she arrived at Sera. The last time she gave birth – at Lewa – she abandoned her calf, so everyone at Sera was on high alert when Lojipu arrived. Sure enough, Nairenyu left Lojipu on his own two days after he was born, so veterinarian Mathew Mutinda swooped in once it became clear that Nairenyu was not going to return.

Mathew administered a mild sedative before loading the calf on to a plane bound for Reteti. There, he met his temporary mother: Mary Lengees, who is not a *Diceros bicornis* but rather a *Homo sapien*. 'I just took Loijipu in and accepted him as a son,' said Mary in a 2020 blog post. 'He used to sleep on my lap, and would cry if we left him. So, I was there, day and night, feeding and comforting him.' Lojipu's wild instincts began kicking in four months later, but he needed additional time to heal and learn. On walks around Reteti, Mary showed the young rhino which plants he could safely eat. Equally important was the care and attention she paid him.

A year and a half later, during the summer of 2018, Lojipu was ready to leave Reteti and return to Sera. Mathew was determined to move the youngster without putting him under anaesthesia since post-translocation complications are more likely to occur following sedation, so the veterinarian and his colleagues patiently waited while Lojipu considered whether to enter his crate. Seven hours later, he voluntarily wandered in. When the Land Cruiser carrying his crate pulled up at Sera, members of the conservancy were there to greet him.

Mary was in tow, as well. She stayed at Sera for two months helping Lojipu adjust, even though she was pregnant at the time. She eventually returned to Reteti. When her daughter was born, Mary decided to name the infant Najipu, which is the feminine counterpart to Lojipu.

With Lojipu back, Sera's black rhino population stood at 13. It has since risen to 19. Sera's rhinos are unique, and not just because they are the only black rhinos living on East African community owned land. Tourists can track them on foot in what tour operator Riccardo Orizo calls 'the most exciting rhino tracking experience in Africa'. (I was thrilled, and more than a little terrified, when I took part in January 2022. My heart nearly thumped out of my chest during the hour I spent watching a male rhino snack on thorn-ridden shrubs from 10 metres away. Rhinos have poor vision, so the male couldn't see my guide nor me, nor the trackers who follow him all day long, but the oxpeckers removing ticks from his back and ears were liable to sound the alarm if they caught a glimpse of us. Fortunately, they kept quiet, and the wind consistently blew towards us, ensuring our scent didn't give us away. We remained undetected long enough for me to take dozens of eye-level photos that will forever remind me that I saw a black rhino in the wild at close enough range that it could have killed me.) Riccardo and his wife, Elizabeth, founded Sera's only safari lodge – Saruni Rhino. Although the Italian duo operate the lodge, the community owns the brick-and-mortar structure. Saruni Rhino, which is a luxury property, opened in January 2017 – *after* rhinos returned to Sera. The timing was intentional: rhinos, the Orizos knew, would draw in tourists. The couple runs four lodges, all in Kenya. Two are in the Maasai Mara, two are up north, and all four properties are located on community lands. When Riccardo and I spoke last summer, I asked him why he and Elizabeth solely operate in community conservancies. Are they guided by a sense of altruism?

Not exactly. Riccardo touched on the importance of community involvement in conservation, saying that Africa's fauna and flora can't be saved unless communities act as environmental guardians, but the decision to open Saruni Rhino was practical. Riccardo sees northern Kenya as the next frontier, 'the future of Kenyan tourism'. He acknowledges, however, that opening a lodge at Sera required taking on risk. 'You have a wonderful community, beautiful sanctuary and the support of a stable organisation like NRT, but we had to ask ourselves, "What if poachers target Sera's rhinos?"' He and Elizabeth knew their reputation would suffer if the project failed, but after asking themselves a string of difficult questions, they decided to go for it.

As it turns out, they bet on a winner. Not only has Sera's rhino population grown, but tourism has risen along with it. According to Reuben, tourism has experienced a 40 per cent annual increase since rhinos were reintroduced in

2015. Unsurprisingly, the pandemic disrupted this growth, but tourism began to rebound in 2021.

In July of last year, the Kenya Wildlife Service released the results of its latest country-wide rhino census. The country's overall rhino population was up: in late 2017, Kenya contained 1,258 rhinos; by mid July 2021, the country housed just under 1,800 individuals. With the exception of northern white rhinos, all of Kenya's rhino subspecies had risen in number, but plans to revive the northern white are rapidly moving forward. In July 2021, the consortium working to save the subspecies announced that three new embryos had been successfully created. They and nine others are on ice in Italy awaiting implantation in southern white surrogates living at Ol Pejeta. Veterinarian and scientist Thomas Hildebrandt, who leads the project, says the first implantation could occur before the end of 2022. In the meantime, black rhinos like Nairenyu (Lojipu's mother) are nurturing the next generation of Kenyan rhinos. The female gave birth to another calf in 2019, and this time, she stuck by its side.

CHAPTER ELEVEN

Recovering the Wild Heart

In 2009, I left a coveted law firm job in the City of London. I was giddy yet nervous: certain I wanted a different career but unsure of its shape. Pivoting within the legal field made the most sense, but deep down, I knew law would end up in my rear-view mirror.

The feeling was uncomfortable. I had wanted to practise law since I was a little girl. In The Forest, I dreamed about exploration and wildlife, but The Forest was a magical realm. I knew that, even as a six-year-old. My professional role models – my father and grandfather – were both corporate attorneys. I was like them: confident, persuasive, an argumentative wordsmith. Law was my destiny.

At university I studied history – the classic American precursor to law – and then I spent a year working at a law firm in London. Then came a three-year juris doctorate at the University of Texas School of Law – the same institution from which my father (and his) graduated. I was over the moon when I landed a job with a top-tier 'Magic Circle' law firm in London upon graduating. There, I earned $160,000 a year providing legal counsel to big banks and mega-corporations, but I felt like a monkey trapped in a cage, not an attorney starting a promising career. My time was not my own, I was chained to my desk and the work bored me to tears. It was time to jump off the legal merry-go-round, but where would I land?

Answering this question required time; I needed space to think. Rather than diving headfirst into another job, I put my career on hold and travelled to Namibia to volunteer at a cattle ranch-turned-wildlife sanctuary. It was my first trip to sub-Saharan Africa and my first time working with wild animals – both dreams come true. Inspired by my surroundings, I began writing. Observations, then fully formed essays, filled my travel journal. Months later, I secured one of my first freelance writing assignments – a digital essay for *The Atlantic* – in which I recounted hearing my first lion roar.[1]

> On its first vocalization, the beautifully intense symphony of roars seemed dangerously close. Sound waves spiraled into the night until they reached the tip of my nose, where they seemed suspended in space and time. Fear took over, conjuring a bone-chilling image: a pride of lions, manes and tawny fur brushing against each other, bounding towards the promise of my human scent.
>
> But the thorn-ridden savanna stood perfectly still. A belated moon rose overhead, casting faint light over a stunning yet uneventful panorama: lions had not escaped their enclosures in a man-hungry frenzy a la *The Ghost and the Darkness*. These were captive lions, marking makeshift territories.

Back then, rewilding was not in my repertoire, but it was all around me as I helped orphaned and injured animals prepare for their return to the wild. In Namibia, I inched towards personal rewilding while experiencing a professional calling that struck me, like a flash of lightning to the soul, during a solitary sunset walk through the bush. I have no interest in speculating about who, or what, made the call. The source, as far as I'm concerned, is irrelevant. What matters is that I was clear-headed enough to receive the message and hungry enough to heed it. Soon after, I walked away from law and began to pursue my two passions: storytelling and wildlife. Before long, I was writing magazine and newspaper articles about conservation and ecotourism, and I began working on a book. It was a memoir of sorts that fused my personal journey with discussions about wildlife conservation. (Sound familiar?)

The working title was *Recovering the Wild Heart*.

Although I shelved it six months later after acknowledging that I wasn't equipped to write a book, the manuscript contained kernels of wisdom, and the title was nothing if not prescient. Chills run down my spine when I think about it now. Without realising it, I had begun rewilding myself; and while I'd never heard the term, I was itching to write about the kinds of projects featured in the preceding chapters. Which isn't to say that I'm a psychic wizard, though I may well be. The point is that on some level, nature wants to call each of us home. We just have to listen.

The decade that followed my trip to Namibia was full of change. Not only did I switch careers, but I moved five times – from London to Washington DC, DC to New York, New York to Texas and Texas (back) to England. In 2015, six years after dipping my toe into the ocean of wildlife conservation, I began a master's degree in conservation leadership at the University of Cambridge. There, I wrote my dissertation on 'conservation storytelling', bringing my previous dream, my calling, full circle. I didn't know it then, but I was already on the path that led to this book. And to a series of milestones that collided in May 2021. I turned 40 while 'celebrating' one year of writing *Wilder*. Like the rest of the world, I continued to grapple with the pandemic's many unwelcome side effects. Although months of lockdown left me feeling as lonely as Richard Henry and Lonesome George combined, I simultaneously yearned for solitude in nature, so I decided to rent a writer's retreat in the English countryside. It turned out to be as productive as it was clichéd – for me personally and for this book. Having spent a year writing about rewilding from urban bases, I needed a quiet, and wilder, place to work.

Oakleigh Cottage is approximately 250 years old. When it was constructed, all of my ancestors lived on this side of the pond, and the United States of America didn't quite exist. Oakleigh's first occupants were employed by a nearby estate, Over Norton Park, which owned the cottage until 1918. Over Norton Park has been in the same family since 1726. Today, two grown siblings, in their 70s and 80s, own the 101-hectare (250-acre) property, which remains a working farm. One of those siblings is evolutionary biologist Richard Dawkins, who is best known for popularising a gene-centric view of evolution (and coining the term *meme*). Dawkins doesn't live on Over Norton Park, but his sister and brother-in-law reside there, as do their adult children.

I didn't know any of this when I signed a lease on Oakleigh Cottage and wasn't looking to find out. My mind was laser-focused on global rewilding, leaving zero bandwidth for local history or new friends. Nature, though, I had time for. Most evenings, I closed my laptop, grabbed my camera and struck off from Oakleigh Cottage towards nearby public footpaths. For the uninitiated, these footpaths are public rights of way that date back to a time when people had to walk across private land to travel between home, work and market towns. Today, they are used recreationally. It didn't take long for me to discover three footpaths within a short distance of my front door. My favourite begins in a field located beside the village's former civic hall. Once you pass through the gate separating private and public property, you slope down a gentle hill that leads to a large pond. Unfortunately, like 99 per cent of the field, it is off limits because the public is only permitted to walk on clearly defined footpaths.

From there, you can either hike up a rocky trail or cut into a field on your right. I explored both routes and later discovered a number of smaller paths splintering off from the two. The various pastures and paths that kept me sane after long days of writing belonged, it seemed, to a single estate – one, I presumed, around which Over Norton was built. In the coming weeks, I explored the village itself while continuing to get my public footpath bearings. In doing so, I came to believe that the entrance to the estate was located just up the street from Oakleigh Cottage.

One sunny afternoon, while heading out on a nature walk, I came upon a couple walking their dog. As soon as I said hello, the young woman (Kitty) asked what an American was doing in the village, and soon we were chattering away. Before long, my book came up. 'Rewilding?' she squealed. 'We are trying to launch a rewilding project down the road!' Although she grew up between the United States and London, her boyfriend (Toby) is from Over Norton, and his family owns a farm on the outskirts of the village. With Kitty's help, Toby hopes to rewild the land. Did I want to come and see it?

As we walked around Toby's family farm, the couple talked about their dreams for it: they want to end all farming operations, restore arable fields and plant native vegetation that will attract wildlife, but first they have to convince the current owners (Toby's parents) that rewilding makes sense. Then they have to iron out the mechanics of rewilding. Although Toby has grown up helping out on the farm, he doesn't have any ecological restoration experience. Kitty, meanwhile, knows a lot about European rewilding – a fellow journalist, she recently wrote an article on Ukrainian rewilding – but she, too, lacks hands-on experience. Whether they pull off their plan remains to be seen, but when we parted ways that sunny Sunday afternoon, we agreed to stay in touch.

A few weeks later, on a sun-less afternoon, I left my cottage and walked up Over Norton's main street. I was planning to cut through the village and walk down my favourite public footpath, but noticed several swallows flying near the stone wall enveloping the estate, so I carried on with hopes of photographing them. As I stood on the pavement with my long lens pointed high in the sky, a man strolling past asked if I was a wildlife photographer. We discussed photography while a light rain began to fall, and then our conversation turned to careers. As fate would have it, my new friend (Craig Blackwell) is an ecologist who has spent decades working on ecological restoration in Oxfordshire. I asked if he knew anyone spearheading rewilding projects in the area.

'Your neighbour is doing a bit of rewilding on his farm,' Craig replied while motioning towards the entrance to the estate. 'He's a retired surgeon by the name of Mike Kettlewell. I don't know him personally, but I'm sure someone in Over Norton can connect you.' Craig confirmed that the property – Over Norton Park – is the estate that employed my cottage's original tenants. And a subsequent Wikipedia search revealed the owners as Richard Dawkins and his sister, Sarah. Mike is Sarah's husband, and he manages environmental efforts on the farm with one of their grown-up sons, Nick (who, I would later learn, runs the farm's commercial operations).

I jotted down Craig's email address, knowing he would be an invaluable resource when I began researching the ecology of the area, and headed home. The coming days were spent trying to figure out how to contact Mike. I hoped that Kitty or Toby could put us in touch, but they had no contact details to share. Since I didn't know anyone else in the village, it became clear that if I wanted to meet Mike I'd have to show up on his doorstep.

Fortunately, the gate was open when I wandered over a few days later, and Mike was standing outside unloading boxes from his truck. He gave me a puzzled look. Even in the countryside, strangers don't typically turn up on one another's doorsteps, and my expat status added an extra layer of self-consciousness. Mike warmed up as soon as I introduced myself and explained that I wanted to learn about the rewilding of Over Norton Park, to which he responded, 'You have the unenviable task of persuading most of the public and politicians that we must rewild.' We drilled down on the point later, but first Mike gave me a tour of the farm.

We hopped in an all-terrain vehicle that looked like a life-size Lego piece and began making our way around Over Norton Park's 166 hectares (410 acres). First up was the sloping field I navigated several times a week. Rolling down it, we came to the pond I always wanted to inspect but hadn't (only a sliver of it is visible from the footpath, and I keep trespassing to a minimum). Mike told me it was a medieval fish pond whose banks now run wild, 'We try to keep it pretty rough, as you can see'. Kingfishers are found there, as are various kinds of waterfowl. I asked Mike if there were any portions of the farm that were left entirely to their own devices. 'The wild bits in the valley,' he responded, 'have never been cultivated.'

Mike proceeded to tell me about a second farm owned by the family that is located several kilometres away and was acquired in the 1980s. On the farm, Mike and Nick have sown wild-flower meadows to attract wildlife and restore plant diversity. Their work began in 1998 – two years before Mike stopped practising medicine – and a second meadow was sown five years ago. Not only are wild-flower meadows rich in plant life – a single one can contain more than 100 wild flower and grass species – they also attract animals, from insects to birds and small mammals. Unfortunately, these biodiverse habitats are in decline. According to Natural England, 97 per cent of the UK's ancient wild-flower meadows disappeared between 1930 and 1980, primarily as a result of agricultural intensification.[2] Having recognised their importance, ecologists are urging landowners to restore forgotten meadows and plant new ones. They emphasise that even small patches in your back garden can make a difference.

Back at Over Norton Park, Mike and I carried on towards one of the cultivated crop fields, this one containing barley, then rolled to a stop and climbed out of the Lego-mobile. Mike wanted to point out two nearby hedgerows that are parallel to each other, making them a 'double hedgerow'. Like wild-flower meadows, hedgerows – which are linear strips of shrubs and trees – offer wildlife food and shelter. Because they typically lie along edges of agricultural lands and roads, hedgerows simultaneously serve as critical wildlife corridors. I can't help but associate hedges with birds, having heard the squeaky calls of countless fledglings tucked within their protective foliage. Farmland birds (which steeply declined across the country when farming intensified) need healthy hedges and scrub to thrive. If properly maintained, hedges provide farm-dwelling birds with nesting places as well as summer and winter food sources, a trio website Hedgelink calls 'the big three'.

Mike and Nick maintain grassy field margins along the edges of all of Over Norton Park's arable fields. These small strips of grass, which can be as narrow as 1m (3.3ft), separate hedgerows and fields. 'Field margins, although generally the least productive areas of a field, can benefit wildlife in many ways,' underscores the Royal Society for the Protection of Birds (RSPB).[3] They attract ground-nesting birds, insects and spiders; as well as small mammals like harvest mice and voles. Mike told me that Over Norton's margins, which are 6m (20ft) wide,* are 'full of butterflies and birds'. Numbers, he adds, have markedly increased since he and his son began restoring biodiversity to parts of the property. When it comes to butterflies, Mike's observations are not yet backed up by data, but birds are another story, as we'll soon see.

They have benefited from ecological restoration, of course, but specialist farmland birds need an extra, and urgent, boost because their numbers have plummeted across the country by around 80 per cent since the early 1970s. Various changes – from a loss of mixed farming to an increase in pesticide use – sparked declines. Many farmers used to leave crop fields fallow during the winter. Because such fields contained crop stubbles and leftover seeds, birds flocked to them. Now, crops are sown as soon as others are harvested. To offset these changes, farmers like Mike create wild bird seed plots next to hedgerows. When plots are depleted (as often happens before the end of winter), farmers can help birds by manually scattering

---

* British farmers participating in land stewardship programmes must adhere to specific requirements. The Kettlewells, for instance, had to keep 6-m (20-ft) wide, florally rich field margins between 1998 and 2021 to qualify for the country's Agri-Environment Scheme (AES). Notably, AES and related initiatives are in the midst of undergoing changes as the Department for Environment, Food and Rural Affairs (Defra) moves away from EU-based rules.

seeds. Supplementary feeding has proven essential since sown seed plots typically become bare during a time of year when wild food sources are scarcest.

Mike and Nick started scattering seeds for wild birds in 2003 and planted their first wild bird seed plot in 2008. When they provide extra food, they do so generously, putting out more seed than is required under British agricultural-environment schemes (more on these schemes in a bit). In recent years, they have collaborated with scientists to assess the impact of their bird-centric initiatives, finding that 'generous daily supplementary feeding' alongside sown seed plots 'reliably attracted and sustained large numbers of priority farmland birds'[4]. During one particularly successful autumn, they recorded more than 500 tree sparrows (the species is on the UK's highest conservation priority list). Mike and his study co-authors were quick to point out, however, that farmers following government feeding protocols 'would be unlikely to meet the evident local demand or the objectives of sustaining sufficient birds to reverse recent trends'. I was beginning to see why Mike called rewilding an uphill battle.

Back at the farmhouse, Mike put the kettle on while I met Sarah and the couple's dog, whose name happens to be Millie (another stroke of awkwardness and/or serendipity). Mike showed me a map of Over Norton Park at its inception, pointing out our location on it, while we made our way to a sunlit conservatory. I asked about Mike and Sarah's early life, learning they were both born in Malawi when it was known as Nyasaland (their fathers worked in colonial agriculture). I wanted to discuss Africa but kept my questions focused, knowing I had imposed enough after showing up on their doorstep.

While sipping tea, we talked about the environmental crisis and the economics of farming. 'Success in the free market is predicated on greed and exploitation,' Mike lamented, 'which explains the state of the planet today.' The environment, he argues, has to compete against the free market. Since most people are financially constrained, the free market model demands cheap food. Someone has to produce it, and farmers have to earn a living. Mike and Sarah can afford to farm less intensively because they have other revenue streams, but Mike believes that if *all* British farmers pursued organic farming and rewilding, the UK would have to import more of its food. We would, in his words, end up 'exporting our problems'.

I knew from Mike's earlier comment that he considered rewilding a constant struggle but was eager to unpack his remark, particularly since Over Norton Park participates in government schemes designed to encourage environmentally friendly practices. As in other countries, the UK compensates farmers who manage their operations in ways that benefit the environment. Many of the efforts described above, from field margins to wild bird seed plots, are covered by agricultural-environment schemes, and Over Norton Park has taken part in them for more than 20 years. Although doing so generates income, they aren't, in Mike's opinion, sufficiently motivating (or particularly user-friendly).

Moreover, as the Over Norton Park bird study demonstrates, wildlife often needs more help than schemes guarantee. The UK stewardship scheme, for instance, funds the scattering of 3.5 tonnes of wild bird seed per year. Mike and Nick end up scattering 4.5 tonnes since, in their experience, 3.5 is not adequate. On top of that, Mike and Nick place artificial nest boxes throughout the farm to support hole-nesting birds like owls and swallows. Craig Blackwell – the ecologist I met walking in the village – agrees that governmental programmes are inadequate, at least when it comes to British farmland birds. 'The proof is in the pudding,' he told me over coffee last August. 'Farmland bird populations are still plummeting across the country despite various agri-environment schemes.'

For 25 years, Craig has witnessed the transformation of Over Norton Park while walking its public footpaths. As an ecologist, he pays close attention to hedge size, tree health, wildlife and more. Over the years, he has seen the farm become 'rougher around the edges and, as a result, more biodiverse'. Among other things, he appreciates that when old trees die on the farm, they are left to rot. 'It looks untidy,' he remarks, 'but is great for invertebrates.' Craig doesn't just appreciate rewilding, however: the retired ecologist has spent decades championing – and in some cases initiating – the practice.

In 1991, upon becoming Oxfordshire's first ecologist, Craig started evaluating the county's biodiversity and developing programmes to improve it. He worked for the county for just under 20 years, during which time he helped initiate a network of Conservation Target Areas that are built around increasing connectivity. Additionally, Craig advised counties on the environmental impact of proposed development while supporting the launch of various community conservation programmes. When he retired, the English ecologist decided to devote his time to smaller-scale projects, several of which are located down the hill from Over Norton in the market town of Chipping Norton. One is a small rewilded plot of land located next to the local health centre. Composed of various plant species that attract bees and butterflies – and bloom at different times of year – it is designed to help pollinators while offering patients a chance to connect with nature. Craig also obtained permission to restore a portion of the local cemetery – an approach that is proving popular across the UK. In Somerset, for instance, 120 parishes have joined a rewilding partnership with the Somerset Wildlife Trust called Wilder Churches.

Micro-rewilding is taking place all over the world. A San Francisco couple cycles around the city with salt-shakers containing wild-flower seeds. They dash seeds on neglected green spaces, like roadside plant patches, in hopes of beautifying the city and, more importantly, supporting pollinators. To get their message across, they often scatter seeds wearing bumblebee costumes. Closer to Over Norton Park are two Staffordshire-based teenagers breeding endangered native reptiles with hopes of releasing them into the wild. London is witnessing its share of active and passive rewilding initiatives, too. During the pandemic, Kensington Gardens (in Hyde Park) converted a manicured stretch of lawn into a wild-flower meadow, and in west London, the Ealing Wildlife Group has begun reintroducing harvest mice to the area. Doing so, conservationists hope, will help bird of prey populations rebound.

In regions like Europe and North America – where rewilding has been mainstreamed – new projects are lifting off practically every day. I know because I signed up for 'rewilding' Google alerts several years ago and have observed a steady increase in media hits over time. I wanted to prove my suspicion (that the word is gaining popularity) without having to review years of emails, so I turned to Google's word history tool, finding that the term's usage slowly grew between the early 1990s and 2010 before experiencing a massive spike during the summer of 2011. It levelled out that August and has since steadily grown at a high rate.

Is this development, however exciting, entirely positive? In the introduction, I mentioned that some conservationists welcome diverse approaches to rewilding while others worry that the term means too many things to mean any one thing. Alongside this concern is another, more serious one. It relates to accountability. How does the rapidly growing rewilding movement ensure that inexperienced individuals and organisations rewild the world in appropriate ways? Should everyone, teenagers and all, feel empowered to rewild even though conservationists can't agree on the meaning of the practice? If individuals choose to rewild, which baselines, if any, should they use? One camp of experts believes baselines should be ignored in rewilding – humans should kick off restoration but let nature decide its own course – while others seek to restore past conditions. And what if wildness, to novices, is an unnatural state? I don't know about you, but I'm guilty of conflating wildness with ecological health.

Across from Oakleigh Cottage is a dilapidated house that has been unoccupied for decades. As a result, the garden is unmanicured, untamed. Put another way, it is, or at least seems, wild. One evening I walked across the street to see more of the property and photograph vines meandering through broken windows. As I stepped off the stone path leading to the front door, my eyes landed on an exquisite oriental poppy in full bloom. The red-and-purple flower (*Papaver orientale*) doesn't belong in the UK. Native to Western Asia and the Caucasus, oriental poppies were introduced in the UK (and elsewhere). Craig tells me they're not considered a pest, but what if they were?

Many of the world's invasive species entered foreign spaces because the introducer believed they contributed to the local environment in some way. Maybe they reminded immigrants of life back home, or perhaps they were, like New Zealand's stoats, seen as solutions to environmental problems. It isn't hard to imagine once-popular acclimatisation societies describing their work in similar terms to rewilding; a member of the American

Acclimatization Society advocated for the introduction of British birds that were 'useful to the farmer and contributed to the beauty of the Groves and Fields'.[5] If he were alive today, he would be horrified to discover that benign-seeming birds like English house sparrows and European starlings had caused significant damage in the United States.

It is unrealistic, I think, to expect the general public to understand the complex distinctions between what *looks* wild and what *is* wild. Axel Moehrenschlager, who chairs the IUCN Conservation Translocation Specialist Group and (briefly) appeared in Chapter Two, worries that people will act on mistaken assumptions. 'If something looks wild, no matter what it is, people might assume it has an ecological benefit.' As a result, they could harm native species by moving or introducing invasive ones.

Even experts make mistakes. Thirty years ago, a group of Finnish scientists introduced a type of caterpillar to an island in the Åland Archipelago without realising that some of the caterpillars had foreign, parasitic wasps living inside them. In the words of a journalist writing for *The Guardian*, the wasps 'sprang out of the butterfly like Russian dolls'.[6] Unfortunately, three decades later, all three species of parasitic wasp continue to live on the island. The tale is cautionary, as Anne Duplouy, one of the Finnish scientists, concedes. 'The reintroduction of endangered species comes from the heart, a good place,' she said, 'but we have a lot to learn about the species we are reintroducing and the habitat where we want to reintroduce them before we do so.'[7] Mistakes like these are costly – particularly when they are irreversible. Damage can be done in indirect ways, as well. In London, park-goers like to feed grey squirrels and ring-necked parrots, both of which are invasive (and, let's be honest, cute). Doing so helps invasive species thrive to the detriment of native ones. Red squirrels, for instance, have significantly decreased in number since grey squirrels arrived from North America.

Another problem stems from the fact that many people have preconceived notions about how nature should feel, look and smell. 'There is often a disconnect between public perceptions of rewilding landscapes and wildlife with what is best ecologically,' wrote several rewilding experts in a book on the subject.[8] They added, 'For example, leaving dead wood in woodlands delivers a variety of important ecological services, including provision of microhabitats, nutrient recycling and carbon storage, but may not always meet with public approval.' The disconnect between what *seems* healthy versus what *is* healthy is only one part of the problem. More often than not, people have a limited tolerance for wildness and all that it brings.

We see this time and again in places where human–wildlife conflict is rife. In places like Africa and India, where people have long coexisted with wildlife, they are willing to tolerate elephants and tigers as long as their safety and livelihoods are protected. Elsewhere, conservationists are struggling to return large mammals (carnivores, in particular) to their native ranges as people push back out of fear – whether for themselves or their livestock. Even smaller-scale initiatives that don't involve carnivore reintroductions are challenged by human attitudes towards nature. In the UK, many landowners have protested beaver reintroductions, fearing crop and tree damage. They don't seem to care that beavers belong in the UK and benefit ecosystems. Either that or their own needs come first.

Human tolerance for wildness has also been put to the test in my home town. When rewilding of the San Antonio River began, people living on properties abutting the Mission Reach were ecstatic about restoration efforts, but their enthusiasm evaporated when they started seeing rodents and snakes on their land as a result of the ecosystem coming back to life. Steven Schauer of the San Antonio River Authority had to meet with residents to explain that these animals are natural and ecologically important, but a fear of rats and snakes doesn't go away overnight. Neither does an ingrained aversion to unmanicured lawns and rotting trees. To truly understand the benefits of rewilding, we have to reprogramme – or *rewild* – ourselves.

Kenya, 2020

The wind whipped through my hair as the open Land Cruiser rolled across one of the Maasai Mara's unpaved roads. I closed my eyes and let the hot sun coat my face in a fiery blaze of orange. Opening my eyes again, I continued scanning the savanna for wildlife. Thirty-five years had passed since I sat in the hunting seat of my grandparents' Chevy Suburban shouting at my dad to drive towards a cluster of scimitar-horned oryx, but the thrill of the game drive remained unchanged. It took several African safaris for me to realise that, in some respects, my trips to the continent are a way of reliving my past. On safari, I can reconnect with the wildness I felt as a young girl while experiencing the world as it used to be, as it should be. Seeing large and potentially dangerous animals like elephants, lions and rhinos humbles the observer. These behemoths remind us that we are small – in physical terms, at least – conjuring a feeling that is uncomfortable yet liberating. A similar effect is produced by looking upon vast landscapes, the Iberás and Maasai Maras of the world. They are wilder worlds, ones that *Homo sapiens* have not managed to fully conquer.

In parallel, writing about other people's rewilding projects allows me to vicariously rewild degraded landscapes while imagining what might have been had my family held on to places like Brady Creek Ranch and Greenwood Farms. I am not prone to bouts of regret, but I spend plenty of time thinking about past forks in the road, *Sliding Doors* moments. If the family hadn't sold the ranch, would I be living in Texas managing a rewilding project instead of writing about conservation from the English countryside? If my father and grandfather had been conservationists instead of lawyers, I might be living in Argentina or Rwanda working as a field biologist. Ultimately, the end result is more important than the paths leading to it. The same can be said of rewilding.

Over Norton, September 2021

Four days before handing in my manuscript, I walked from Oakleigh Cottage to Over Norton Park. It was a spectacular early autumn day, one of the prettiest I've seen during 10 years in England. A bright sun warmed my neck as I approached the entrance to my favourite public footpath. As I strolled down the sloping hill towards the medieval fish pond that Mike keeps rough around the edges, I heard the unmistakeable call of a red kite. Looking up, I watched the fork-tailed raptor gliding through the skies. The beautiful bird has experienced a rewilding of its own. Driven to extinction in England (in 1871), red kites were reintroduced in 1990 when 13 individuals were transported from Spain to the Chiltern Hills, a chalk escarpment halfway between London and Over Norton. Today, 2,000 breeding pairs occupy England; the bird of prey is found in almost every English county. Seeing one hunt on Over Norton Park brought a smile to my face, but it faded as I carried on walking.

I was extremely distracted. For weeks, I had been experiencing something akin to writer's block as I attempted to end this chapter and, thus, this book. It wasn't that I couldn't find words. I summoned plenty – too many, in fact. For days, I drafted fresh material only to toss it aside hours later. One section focused on legacy in the conservation context, ending with the point that, by funding large-scale, high-profile projects, philanthropists achieve a kind of immortality. Another day, I spent hours detailing my brush with rewilding. Seven years ago, I nearly purchased what remained of Greenwood Farms with hopes of restoring the old farmhouse and the land on which it sits, but I was admitted to a conservation master's programme at Cambridge the same month I considered making an offer on the crumbling farmhouse. I followed my gut and moved back to England even though it meant saying goodbye to Greenwood Farms once and for all. I scrapped other sections, too, like one on two Ugandan men I met in January 2020 who want to rewild a crater lake beside their village by fencing out cattle.

All this hemming and hawing reminded me that rewilding is happening all around us, all the time, in diverse manifestations. There are, simply put, too many examples to mention in a single book. Which explains why I felt the need to end *Wilder* with an impeccable example of the practice, one that encapsulates all that rewilding stands for, from trophic complexity to human rewilding. That goal, I realised, was unreasonable. The beauty of rewilding is that it means different things to different people and takes a dazzling array of shapes.

Back in Over Norton Park, I slipped through a cattle-proof gate and began walking across the open field where the red kite caught my eye an hour earlier. I was following the footpath across the field, my ears primed for birdsong, when I experienced a lightbulb moment. What I realised is this: the diversity of approach and perspective that defines the practice of rewilding likewise applies to how *we* contribute to environmental protection – and how *we* rewild ourselves. Like keystone species on the food web, powerful, resource-rich people and organisations can rewild in ways that impact entire ecosystems, but members of the public have important parts to play. Whether exerting bottom-up influence like freshwater mussels or engineering creative solutions in the spirit of ecosystem architects, each of us is capable of making the world a wilder, healthier place.

How we do so is up to us.

Acknowledgements

Wilder required input and support from dozens of conservationists and rewilders. I am incredibly grateful to the many individuals who appeared in the preceding pages, all of whom set aside time to patiently walk me through their work. Their generosity – of energy and expertise – was invaluable. Other experts guided me through the fundamentals of rewilding while helping me identify global projects. I would like to thank, in particular, David Banks, Steve Carver, Ian Convery, Caroline Lees, Phil Miller, Axel Moehrenschlager, Nathalie Pettorelli and John Terborgh.

I greatly benefited from the conservation network associated with the University of Cambridge MPhil in conservation leadership programme, through which critical beta readers emerged. Thanks go to Pepe Clark, Daniel Flenley and Noa Steiner for reviewing chapters. I am also indebted to Isabel Vique Bosquet, Leonor Fishman, Sorosh Poya-Faryabi and Carolina Proaño-Castro of my cohort for participating in fascinating (and productive) brainstorms.

I could not have gotten *Wilder* over the finish line without research support from Charlotte Bird and William Byrd. Their ability to rapidly gather and organise materials was unmatched, and they gamely provided feedback on numerous chapters, too. Not only that, but they became critical sounding boards as time went on.

I have wonderful friends who have supported me (and my career) for years. When it comes to *Wilder*, Laura Dodd, Liza Binkley, Mollie Binkley and Jordana Krohley went above and beyond, as did as did Susana Canseco and Brandon Seale.

My team at Bloomsbury has been supportive from the start. Many thanks go to Jim Martin for taking an interest in *Wilder* and the projects it covers. His enthusiasm for rewilding helped fuel my writing, and I appreciate his thoughtful input throughout the research phase. I am grateful to my editor, Angelique Neumann, as well as Allison Davis, Amy Greaves and Robin Wane.

Writing *Wilder* was a largely solitary effort undertaken during peculiar pandemic times. A small circle of family members, friends and pets made the two-year effort not only tolerable but (relatively) enjoyable. I am particularly grateful to my parents, John and Susan Kerr, for welcoming me into to my childhood home for seven months of the pandemic. Their company was indispensable, as was the input they provided as I navigated the colossal task of book-writing. More broadly, they have always supported my professional ambitions even when I made radical changes others couldn't understand. My brothers and their partners were also wonderful cheerleaders, and I owe special thanks to Jeff for critiquing several of my chapters. I knew I was on track when he complimented my writing.

Above all, I am grateful to my parents and grandparents for facilitating a childhood full of creative play, storytelling and time outdoors – experiences that made me who I am. I was further influenced by professional nature storytellers (David Attenborough, Jane Goodall, A. A. Milne and Beatrix Potter to name a few) and teachers who encouraged my creative side (Sharon Moa, Joane Cox and Barbara Bennett).

Notes

Introduction

1. Soulé, M. and Noss, R. 1998. Rewilding and biodiversity: complementary goals for continental conservation. *Wild Earth* 8: 19–28.
2. Perino, A. *et al.* 2019. Rewilding complex ecosystems. *Science* 364(6438): 1–9.
3. Hayward, M. *et al.* 2019. Reintroducing rewilding to restoration – rejecting the search for novelty. *Biological Conservation* 233: 255–9. 10.1016/j.biocon.2019.03.011.
4. Holmes, G. *et al.* 2020. What is rewilding, how should it be done, and why? A Q-method study of the views held by European rewilding advocates. *Conservation and Society* 18(2): 77–88.
5. Jørgensen, D. 2014. Rethinking rewilding. Geoforum. https://doi.org/10.1016/j.geoforum.2014.11.016.

Chapter One

1. Duke, C. in Hanes, S. 2007. Greg Carr's big gamble. *Smithsonian Magazine*.
2. Hatton, J., Couto, M. and Oglethorpe, J. 2001. *Biodiversity and War: A Case Study of Mozambique*. Biodiversity Support Program, Washington, DC.
3. Carroll, S. B. 2016. Resurrecting Mozambique's magnificent Gorongosa. *Sierra*.
4. Hanes, S. 2007. Greg Carr's big gamble. *Smithsonian Magazine*.
5. Herrero, H. *et al.* 2020. A healthy park needs healthy vegetation: the story of Gorongosa National Park in the 21st century. *Remote Sensing* 12(476): 1–23.
6. Treuer, D. 2021. Return the national parks to the tribes. *The Atlantic*.
7. Keeping the mountain healthy will be increasingly important as the detrimental effects of climate change increase, as

Mozambicans well know. Because of the country's location and long coastline, it is incredibly vulnerable to storms that roll in from the Indian Ocean. In 2019, Cyclone Idai ravaged Mozambique, killing 600 people – and leaving hundreds of thousands homeless – while causing severe flooding in Gorongosa and adjoining areas. Although scientists have not proven links between climate change and Idai's intensity, Mozambique is undoubtedly vulnerable to the effects of climate change (according to the Global Climate Risk Index, Mozambique was the fifth-most impacted country in the world between 2000 and 2019).

8. Hoffner, E. 2020. Gorongosa National Park is being reforested via coffee and agroforestry. *Mongabay*.

9. Woodroffe, R. and Sillero-Zubiri, C. 2020. *Lycaon pictus* (amended version of 2012 assessment). The IUCN Red List of Threatened Species 2020. e.T12436A166502262.

10. Weiss, S. 2020. The wild experiment to bring apex predators back from the brink. *Wired*.

11. Daskin, J. H., Stalmans, M. and Pringle, R. M. 2016. Ecological legacies of civil war: 35-year increase in savanna tree cover following wholesale large-mammal declines. *Journal of Ecology* 104(1): 79–89.

12. BBC News (2015). Mozambique declared free of landmines. *BBC News*.

13. Stalmans, M. and Margarida, V. 2018. Forest Cover on Gorongosa Mountain. *Parque Nacional da Gorongosa* 1–14.

14. Kerr, M. 2020. Private conservation. *Delta Sky*.

Chapter Two

1. Aitken, R. B. 1969. *Great Game Animals of the World*. Macmillan Company, New York.

2. Ibid.

3. Ibid.

4. Attenborough, D. 2017. *Adventures of a Young Naturalist: the ZooQuest Expeditions*. Two Roads, London.

5. The Antelope Collectors in *Zoo Studies*. 2019. McGill–Queen's University Press, Kingston, Ontario 45–64. Referring to Carl Hagenbeck to William T. Hornaday, 28 August 1903,

New York Zoological Park, Office of Director, William T. Hornaday and W. Reid Blair Incoming Correspondence and Subject Files 1895–1940. The wavy underline appears to be in Hornaday's hand.

6. The Antelope Collectors in *Zoo Studies*. 2019. McGill-Queen's University Press, Kingston, Ontario 45–64. Referring to William T. Hornaday's National Collection of Heads and Horns in Zoological Society Bulletin.

7. Seddon, P. J., Armstrong, D. P. and Maloney, R. F. 2007. Developing the science of reintroduction biology. *Conservation Biology* 21(2): 303–12.

8. Knight, T. and Rose, M. 2017. *With Honourable Intent*. William Collins, Scotland.

9. Woodfine, T. and Gilbert, T. 2016. The fall and rise of the scimitar-horned oryx in *Antelope Conservation: from diagnosis to action*. John Wiley & Sons, New Jersey.

Chapter Three

1. Clark, A. 1949. The invasion of New Zealand by people, plants and animals. *The South Island*. Greenwood Press, Connecticut.

2. Latham, A. D. M. *et al.* 2020. A refined model of body mass and population density in flightless birds reconciles extreme bimodal population estimates for extinct moa. *Ecography* 43(3): 353–64.

3. King, M. 2012. *The Penguin History of New Zealand*. Read How You Want, Sydney.

4. Stevens, W. 2000. Suspects in 'Blitzkrieg' extinctions: primitive hunters. *The New York Times*.

5. Clark, A. 1949. The invasion of New Zealand by people, plants and animals. *The South Island*. Greenwood Press, Connecticut.

6. Forest and Bird. 2018. *Frequently Asked Questions about 1080*.

7. Russell, J. C. *et al.* 2015. Predator-free New Zealand: conservation country. *BioScience* 65(5): 520–5.

8. King, M. 2012. *The Penguin History of New Zealand*. Read How You Want, Sydney.

9. Ibid.

10. Wells, P. 2006. 'An enemy of the rabbit': the social context of acclimatisation of an immigrant killer. *Environment and History* 12(3): 297–324.

11. Darwin, C. 1859. *The Origin of Species by Means of Natural Selection.* 6th edition. Josh Murray, London.

12. Hill, J. 2021. *Life of Richard Henry.* Unpublished document.

13. Hill, J. 2021. *Richard Henry.* Message to Millie Kerr (email).

14. Henry, R. 1907. Small birds and kākāpō – natives competing with the imported. *Richard Henry of Resolution Island.* Cadsonbury Publications, New Zealand 12–13.

15. Hill, J. 2021. *Life of Richard Henry.* Personal communication.

16. Merton, D. 1987. *Richard Henry of Resolution Island.* Cadsonbury Publications, New Zealand.

17. Otago Daily Times. 1949. Gained in telling. Papers Past.

18. Merton, D. 1987. *Richard Henry of Resolution Island.* Cadsonbury Publications, New Zealand.

19. Ibid.

20. SPCA, 2019. *1080 – what is it, and what can be done about it?*

21. Forest and Bird. 2018. *Frequently Asked Questions about 1080.*

22. Tompkins, D. 2021. New Zealand's mission to eradicate key threats to biodiversity. Predator Free 2050 PowerPoint.

23. Russell, J. and Malhi, Y. 2019. Comment. In response to Perino, A. *et al.* 2019. Rewilding complex ecosystems. *Science* 364(6438): 1–9.

24. Ibid.

25. Russell, J. C. *et al.* 2015. Predator-free New Zealand: conservation country. *BioScience* 65(5): 520–5.

26. Merton, D. 1987. *Richard Henry of Resolution Island.* Cadsonbury Publications, New Zealand.

27. Otago Witness. 1896. Resolution Island: the sanctuary for wingless birds. Papers Past.

Chapter Four

1. Pietersen, D. and Connelly, E. 2019. *Smutsia temminckii.* The IUCN Red List of Threatened Species 2019: e.T12765A123585768.

2. Conciatore, J. 2019. Up to 2.7 million pangolins are poached every year for scales and meat. awf.org.

Chapter Five

1. Butler, S. 1898. *The Iliad of Homer*. Longmans, Green and Co. Rendered into English prose. Tufts University Perseus Digital Library.
2. Macro Trends. *San Antonio Metro Area Population 1950–2021*. macrotrends.net.
3. Ibid.
4. Dukes, T. 2021. Phil Hardberger park geology. *Texas Master Naturalist*.
5. Patoski, J. 1997. The war on cedar. *Texas Monthly*.
6. Google. Google Earth – Timelapse.
7. Noonan Guerra, M. A. 1987. Excerpt from *The San Antonio River*. University of Incarnate Word.
8. Ibid.
9. Miller, C. 2012. *Deep in the heart of San Antonio: land and life in South Texas*. Trinity University Press, Texas.
10. Weather and Climate. 2021. Average monthly snow and rainfall in San Antonio (Texas) in inches. weather-and-climate .com.
11. Climate Data. 2021. Climate London (United Kingdom). en.climate-data.org.
12. Eckhardt, G. Regional climate and water availability. edwardsaquifer.net.
13. White, T. 2018. Record-setting flood of 1998 happened 20 years ago this week in San Antonio area, South Texas. mysanantonio.com.
14. Ibid.
15. Carnett, L. 2018. SA gears: a look inside Olmos Dam. mysanantonio.com.
16. Texas Beyond History. Olmos Dam. texasbeyondhistory.net.
17. Scott, S. 2013. The Mission Reach: bringing life and pride back to the southside. *San Antonio Report*.
18. The Living New Deal. River Walk – San Antonio TX. livingnewdeal.org.

19. Olmsted, F. L. 1857. A journey through Texas or a saddle-trip in the southwestern frontier. Independently published.
20. Pacific Northwest Freshwater Mussel Workgroup. About Freshwater Mussels. pnwmussels.org.
21. The San Antonio River Authority. The San Antonio River Improvements Project. pnwmussels.org.
22. UNESCO. San Antonio Missions. whc.unesco.org.

Chapter Six

1. Human Rights Watch. 1992. Talking peace and waging war – human rights since the October 1990 invasion. *Human Rights Watch* 4(3).
2. Panthera. Lions. panthera.org.
3. World Wildlife Fund. Tigers. wwf.org.
4. African Parks Network. Akagera National Park. africanparks.org.
5. Gourevitch, P. 1999. *We Wish to Inform You That Tomorrow We Will Be Killed With Our Families.* Picador, London.
6. Peace Agreement between the government of the Republic of Rwanda and the Rwandese Patriotic Front. 1993. Ulster University International Conflict Resource Centre.
7. Bouché, P. 1998. *Les aires protégées dans la tourmente. Evolution de la situation de 1990 à 1996. Cahiers d'Ethologie* 18(2): 161–74.
8. Ibid.
9. African Parks Network. *Aerial Survey Report: Akagera, 2013.*
10. Wronski, T. *et al.* 2017. Pastoralism versus agriculturalism – how do altered land-use forms affect the spread of invasive plants in the degraded mutara rangelands of north-eastern Rwanda? *Plants* 6(2): 19.
11. Bouché, P. 1998. *Les aires protégées dans la tourmente. Evolution de la situation de 1990 à 1996. Cahiers d'Ethologie* 18(2): 161–74.
12. William, S. D. and Ntayombya, P. 1999. Akagera Biodiversity Project final report. Zoological Society of London, UK, and Ministry of Agriculture, Livestock, Environment & Rural Development, Rwanda.
13. Speke, J. in Hanning, J. (1868), *Journal of the Discovery of the Source of the Nile,* Harper & Bros London.
14. African Parks Network. Akagera National Park. africanparks.org.

Chapter Seven

1. Fundación Rewilding Argentina. 2018. *Rewilding Iberá* documentary. YouTube.
2. National Park Service. 2018. Yellowstone National Park Protection Act 1872. nps.gov.
3. Ibid.
4. Doherty, B. 2020. Icon: Rewilding our parks, our hearts, and our wallets. *Forbes*.
5. Zaitchik, A. 2018. How conservation became colonialism. *Foreign Policy*.
6. Heinonen, S. 14 April 2021. Personal interview.
7. Tompkins Conservation. Our work – Argentina. tompkinsconservation.org.
8. Ibid.
9. Heinonen, S. 14 April 2021. Personal interview.
10. Fundación Rewilding Argentina. 2018. *Rewilding Iberá* documentary. YouTube.
11. Van Ausdal, S. and Wilcox, R. W. 2013. Hoofprints: ranching and landscape transformation. *RCC Perspectives* 7: 73–80.
12. BBC News. 2019. Argentina profile – timeline. *BBC News*.
13. Muzzio, M. *et al.* 2018. Population structure in Argentina. *PLoS ONE* 13(5): e0196325.
14. Zamboni, T., Di Martino, S. and Jiménez-Pérez, I. 2017. A review of a multispecies reintroduction to restore a large ecosystem: the Iberá Rewilding Program (Argentina). *Perspectives in Ecology and Conservation* 15(4): 248–56.
15. Burkart, R. 1990. Use and management of natural resources in Argentina's protected areas. fao.org.
16. Zamboni, T., Di Martino, S. and Jiménez-Pérez, I. 2017. A review of a multispecies reintroduction to restore a large ecosystem: the Iberá Rewilding Program (Argentina). *Perspectives in Ecology and Conservation* 15(4): 248–56.
17. Heinonen, S. 14 April 2021. Personal interview.
18. Politi, D. 2020. Argentina teeters on default, again, as pandemic guts economy. *The New York Times*.
19. Conservation Land Trust Argentina. Project Iberá – Giant anteater. proyectoiberá.org.

20. Tapir Specialist Group. What is a Tapir? tapirs.org.
21. Naveda, A. *et al.* 2008. *Tapirus terrestris*. The IUCN Red List of Threatened Species 2008. e.T21474A9285933.
22. Tapir Specialist Group. What is a Tapir? tapirs.org.
23. Ibid.
24. Fundación Rewilding Argentina. Green-winged macaw. proyectoiberá.org.
25. Zamboni, T., Di Martino, S. and Jiménez-Pérez, I. 2017. A review of a multispecies reintroduction to restore a large ecosystem: the Iberá Rewilding Program (Argentina). *Perspectives in Ecology and Conservation* 15(4): 248–56.
26. Grajal, A. and Sanz D'Angelo, V. 2002. Reintroduction of captive raised *Amazona barbadensis* in Venezuela: a case study with a review of guidelines for parrot reintroduction programs. V International Parrot Convention, Spain, 18–21 September.
27. Estes, J. A. *et al.* 2011. Trophic downgrading of planet Earth. *Science* 333(6040): 301–6.
28. Terborgh, J. *et al.* 2001. Ecological meltdown in predator-free forest fragments. *Science* 294(5548): 1923–6.
29. Sanderson, E. W. *et al.* 2021. The case for reintroduction: the jaguar (*Panthera onca*) in the United States as a model. *Conservation Science and Practice* 3(6): e392.
30. van Maanen, E. and Convery, I. 2016. Rewilding: The realisation and reality of a new challenge for nature in the twenty-first century in *Changing Perceptions of Nature*. Boydell & Brewer, Newcastle.
31. Center for Biological Diversity. Natural history: jaguar. biological diversity.org.
32. Caruso, F. and Jiménez-Pérez, I. 2013. Tourism, local pride, and attitudes towards the reintroduction of a large predator, the jaguar *Panthera onca* in Corrientes, Argentina. *Endangered Species Research* 21: 263–72.
33. Ibid.
34. Ibid.
35. Ibid.
36. Tompkins Conservation. 2020. Restoring the Gran Chaco's most iconic species.
    Tompkins Conservation press release sent via email.

37. Secretariat of the Convention on Biological Diversity. 2020. Global Biodiversity outlook 5 – summary for policy-makers. Montréal. cbd.int.

Chapter Eight

1. Darwin, C. 1859. *On the Origin of Species*. John Murray, London.
2. Beer, G. 1996. Introduction in *On the Origin of Species*. Oxford University Press, Oxford.
3. Darwin, C. 1959. *The Voyage of the Beagle*. Dutton, London.
4. Blake, S. *et al.* 2012. Seed dispersal by Galápagos tortoises. *Journal of Biogeography* 39(11): 1961–72.
5. Watson, J. *et al.* 2010. Mapping terrestrial anthropogenic degradation on the inhabited islands of the Galápagos Archipelago. *Oryx* 44(1): 79–82.
6. Nicholls, H. 2014. *The Galápagos: a natural history*. Basic Books, New York.
7. BBC News. 2020. Diego, the Galápagos tortoise with a species-saving sex drive, retires. *BBC News*.
8. Galápagos Conservancy. Project Isabela. galápagos.org.
9. Caccone, A. *et al.* 1999. Origin and evolutionary relationships of giant Galápagos tortoises. *Proceedings of the National Academy of Sciences of the United States of America* 96(23): 13223–8.
10. Jepson, P. and Blythe, C. 2020. *Rewilding: The Radical New Science of Ecological Discovery*. Icon Books, London.
11. Safford, R. Important bird areas in Africa and associated islands – Mauritius. *BirdLife International*.
12. Gibbs, J. P. *et al.* 2014. Demographic outcomes and ecosystem implications of giant tortoise reintroduction to Española Island, Galápagos. *PLoS ONE* 9(11): e110742.
13. Tapia, W. H. 2015. Expedition to Wolf Volcano to search for tortoises. galápagos.org.

Chapter Nine

1. Fernando, P., Leimgruber, P., Prasad, T. and Pastorini, J. 2012. Problem-elephant translocation: translocating the problem and the elephant? *PLoS ONE* 7(12): e50917.

2. Baskaran, N. 1998. Ranging and resource utilization by Asian elephant (*Elephas maximus Lin.*) in Nilgiri Biosphere Reserve, South India. Ph.D. thesis, Bharathidasan University, Thiruchirapally.
3. Ibid.
4. IUCN African Elephant Specialist Group. African elephants – frequently asked questions. iucn.org.
5. Times of India. 2017. By 2030, world will have 8.6 billion people, 1.5 billion of them in India. *Times of India*. and United Nations, Department of Economic and Social Affairs, Population Division (2019). World Population Prospects 2019: Highlights (ST/ESA/SER.A/423). population.un.org.
6. Dutt, B. 2019. *Rewilding India*. Oxford University Press, Oxford.
7. Anandakrishna, K. 2010. The cruelty called kheddas. *The Hindu*.
8. Kumar, R. 2020. Elephants' migration to Kabini may be delayed this year. *The Hindu*.
9. Vidya, T. N. C. and Thuppil, V. 2010. Immediate behavioural responses of humans and Asian elephants in the context of road traffic in southern India. *Biological Conservation* 143(8): 1891–1900.
10. Wadey, J. *et al.* 2018. Why did the elephant cross the road? The complex response of wild elephants to a major road in Peninsular Malaysia. *Biological Conservation* 218: 91–8.
11. Menon, V. *et al.* 2017. *Right of Passage: Elephant Corridors of India*. Conservation Reference Series No. 3. Wildlife Trust of India, New Delhi.
12. Wanli, Y. and Yingqing, L. 2021. Elephants return from epic journey. *China Daily*.

Chapter Ten

1. Jebet, V. 2020. Spotlight on Sera for World Rhino Day. nrt-kenya.org.
2. Levikov, N. 2015. Will community management help save the black rhino from extinction? *Mongabay*.
3. Vitale, A. 2015. Pictures: Black rhinos back in tribal Africa after 25-year absence. *National Geographic*.

4. Tompkins Conservation. *Our work – Argentina*. tompkinsconservation.org.
5. Odote, C. 2013. The dawn of Uhuru? Implications of constitutional recognition of communal land rights in pastoral areuyas of Kenya. *Nomadic Peoples* 17(1): 87–105.
6. Kenya Wildlife Service. 2012. Conservation and Management Strategy for the Black Rhino (*D. b. michaeli*) in Kenya. Kenya Wildlife Service, Nairobi.
7. Jebet, V. 2020. Spotlight on Sera for World Rhino Day. nrt-kenya.org.
8. Kenyan Wildlife Conservancies Association. 2019. *KWCA Action Plan 2009–2013*. kwcakenya.com.
9. Ogutu J. O. *et al*. 2016. Extreme wildlife declines and concurrent increase in livestock numbers in Kenya: what are the causes? *PLoS ONE* 11(9): e0163249.
10. Northern Rangelands Trust. *Who we are*. nrt-kenya.org.
11. IUCN/SSC. 2013. Guidelines for reintroductions and other conservation translocations. Version 1.0. Gland, Switzerland: IUCN Species Survival Commission, viiii 57.
12. Haines, G. 2018. Gunfights, cow theft and love songs: the 'singing wells' of Namunyak. adventure.com.
13. Lowaktuk, D. in *Shaba*. Directed by Vitale, A. 2021. Passion Planet.

Chapter Eleven

1. Kerr, M. 2011. Wild voluntourism: one story of working with big cats in Namibia. *The Atlantic*.
2. Cooke, A. 2017. How to create a wild-flower meadow. naturalengland.blog.gov.uk.
   Gillies, J. 2018. Wild-flower meadows for all. insideecology.com.
3. RSPB. Wild bird seed mixtures. rspb.org.uk.
4. Broughton, R. K. *et al*. 2020. Intensive supplementary feeding improves the performance of wild bird seed plots in provisioning farmland birds throughout the winter: a case study in lowland England. *Bird Study* 67(4): 409–19.
5. 1877. American Acclimatization Society. *The New York Times*.

6. Barkham, P. 2021. Butterflies released in Finland contained parasitic wasps – with more wasps inside. *The Guardian*.
7. Ibid.
8. Durant, S., Pettorelli, N. and Du Toit, J. The future of rewilding: fostering nature and people in a changing world in *Rewilding*. Cambridge University Press, Cambridge.

Select Bibliography

Introduction

Hayward, M. W. *et al.* 2019. Reintroducing rewilding to restoration – rejecting the search for novelty. *Biological Conservation* 233: 255–9.

Chapter One

De Merode, E. *et al.* 2007. The impact of armed conflict on protected-area efficacy in Central Africa. *Biology Letters* 3(3): 299–301.

Eckstein, D., Künzel, V. and Schäfer, L. 2021. Global Climate Risk Index 2021.

Gaynor, K. M. *et al.* 2016. War and wildlife: linking armed conflict to conservation. *Frontiers in Ecology and the Environment* 14(10): 533–42.

Guyton, J. A. *et al.* 2020. Trophic rewilding revives biotic resistance to shrub invasion. *Nature Ecology and Evolution* 4: 712–24.

Hayward, M. W. *et al.* 2006. Prey preferences of the African wild dog *Lycaon pictus* (Canidae: Carnivora): ecological requirements for conservation. *Journal of Mammalogy* 87(6): 1122–31.

Lynam, T., Cunliffe, R. and Mapaure, I. 2004. Assessing the importance of woodland landscape locations for both local communities and conservation in Gorongosa and Muanza districts, Sofala province, Mozambique. *Ecology and Society* 9(4): 1–24.

Rangarajan, M. 2003. Parks, politics and history: conservation dilemmas in Africa. *Conservation and Society* 1(1): 77–98.

Rothfels, N. 2019. The Antelope Collectors. *Zoo Studies: A New Humanities.* Ed. T. McDonald and D. Vandersommers. McGill-Queen's University Press, Montreal 45–64.

Santos, M., Ferreira, A. and Gardete, D. 2020. Constructive characterization of Gorongosa National Park Villages. *Knowledge E Engineering* 2020: 187–201.

Saunders, D. A., Hobbs, R. J. and Margules, C. R. 1991. Biological consequences of ecosystem fragmentation: a review. *Conservation Biology* 5(1): 18–32.

Unruh, J. and Bailey, J. 2009. Management of spatially extensive natural resources in postwar contexts: working with the peace process. *GeoJournal* 74(2): 159–73.

Van Aarde, R., Whyte, I. and Pimm, S. L. 1999. Culling and the dynamics of the Kruger National Park African elephant population. *Animal Conservation* 2(4): 287–94.

Chapter Two

Brugière, D. and Scholte, P. 2013. Biodiversity gap analysis of the protected area system in poorly documented Chad. *Journal for Nature Conservation* 21(5): 286–93.

Durant, S. M. *et al.* 2014. Fiddling in biodiversity hotspots while deserts burn? Collapse of the Sahara's megafauna. *Diversity and Distributions* 20(1/2): 114–22.

Iyengar, A. *et al.* 2007. Remnants of ancient genetic diversity preserved within captive groups of scimitar-horned oryx (*Oryx dammah*). *Molecular Ecology* 16(12): 2436–49.

Kitagawa, C. 2008. On the presence of deer in Ancient Egypt: analysis of the osteological record. *The Journal of Egyptian Archaeology* 94: 209–22.

Ogden, R. *et al.* 2020. Benefits and pitfalls of captive conservation genetic management: evaluating diversity in scimitar-horned oryx to support reintroduction planning. *Biological Conservation* 241: 1–11.

Soulé, M. *et al.* 1986. The millenium ark: how long a voyage, how many staterooms, how many passengers? *Zoo Biology* 5(2): 101–13.

Stanley Price, M. 2016. Reintroduction as an antelope conservation solution in *Antelope Conservation: from diagnosis to action*. John Wiley & Sons, New Jersey.

Chapter Three

Burns, B., Innes, J. and Day, T. 2012. The use and potential of pest-proof fencing for ecosystem restoration and fauna conservation in New Zealand, in *Fencing for Conservation: restriction of evolutionary potential or a riposte to threatening processes?* Springer, New York 65–90.

Ceballos, G., Ehrlich, P. R. and Dirzo, R. 2017. Biological annihilation via the ongoing sixth mass extinction signaled by vertebrate population losses and declines. *Proceedings of the National Academy of Sciences of the United States of America* 114(30): E6089–96.

Cooper, A. and Cooper, R. A. 1995. The Oligocene bottleneck and New Zealand biota: genetic record of a past environmental crisis. *Proceedings: Biological Sciences* 261(1362): 293–302.

Duncan, R. P. and Blackburn, T. M. 2004. Extinction and endemism in the New Zealand Avifauna. *Global Ecology and Biogeography* 13(6): 509–17.

Elliott, G. *et al.* 2010. Stoat invasion, eradication and re-invasion of islands in Fiordland. *New Zealand Journal of Zoology* 37(1): 1–12.

Fisher, D. O. 2011. Trajectories from extinction: where are missing mammals rediscovered? *Global Ecology and Biogeography* 20(3): 415–25.

Maxwell, J. M. and Jamieson, I. G. 1997. Survival and recruitment of captive-reared and wild-reared takahē in Fiordland, New Zealand. *Conservation Biology* 11(3): 683–91.

McLoughlin, S. 2001. The breakup history of Gondwana and its impact on pre-Cenozoic floristic provincialism. *Australian Journal of Botany* 49(3): 271–300.

Oppel, S. *et al.* 2014. Habitat-specific effectiveness of feral cat control for the conservation of an endemic ground-nesting bird species. *Journal of Applied Ecology* 51(5): 1246–54.

Tennyson, A. J. D. 2010. The Origin and History of New Zealand's Terrestrial Vertebrates. *New Zealand Journal of Ecology* 34(1): 6–27.

Chapter Four

Cooney, R. *et al.* 2016. From poachers to protectors: engaging local communities in solutions to illegal wildlife trade. *Conservation Letters* 10(3): 367–74.

Mahmood, T. *et al.* 2019. Distribution and illegal killing of the endangered Indian pangolin *Manis crassicaudata* on the Potohar Plateau, Pakistan. *Oryx* 53(1): 159–64.

Mambeya, M. M. *et al.* 2018. The emergence of a commercial trade in pangolins from Gabon. *African Journal of Ecology* 56(3): 601–9.

Shairp, R. *et al.* 2016. Understanding urban demand for wild meat in Vietnam: implications for conservation actions. *PLoS ONE* 11(1): e0134787.

Shepherd, C. R *et al.* 2017. Taking a stand against illegal wildlife trade: the Zimbabwean approach to pangolin conservation. *Oryx* 51(2): 280–5.

Sun, N. C., Pei, K. J. and Lin, J. 2019. Attaching tracking devices to pangolins: A comprehensive case study of Chinese pangolin *Manis pentadactyla* from southeastern Taiwan. *Global Ecology and Conservation* 20(e00700): 1–8.

Yang, L. *et al.* 2018. Historical data for conservation: reconstructing long-term range changes of Chinese pangolin (*Manis pentadactyla*) in eastern China (1970–2016). *Proceedings of the Royal Society B* 285: 20181084.

Zhang, F. *et al.* 2019. Halting the release of the pangolin *Manis javanica* in China. *Oryx* 53(3): 411–12.

Chapter Five

Vaughn, C. C., Nichols, S. J. and Spooner, D. E. 2008. Community and foodweb ecology of freshwater mussels. *Journal of the North American Benthological Society* 27(2): 409–23.

Chapter Six

Bariyanga, J. D. *et al.* 2016. Effectiveness of electro-fencing for restricting the ranging behaviour of wildlife: a case study in the degazetted parts of Akagera National Park. *African Zoology* 51(4): 183–91.

Rahmanian, S. *et al.* 2019. Effects of livestock grazing on soil, plant functional diversity, and ecological traits vary between regions with different climates in northeastern Iran. *Ecology and Evolution* 9(14): 8225–37.

Rutagarama, E. and Martin, A. 2006. Partnerships for protected area conservation in Rwanda. *The Geographical Journal* 172(4): 291–305.

Saunders, D. A., Hobbs, R. J. and Margules, C. R. 1991. Biological consequences of ecosystem fragmentation: a review. *Conservation Biology* 5(1): 18–32.

Spinage, C. A. and Guinness, F. E. 1971. Tree survival in the absence of elephants in the Akagera National Park, Rwanda. *Journal of Applied Ecology* 8(3): 723–8.

Unruh, J. and Bailey, J. 2009. Management of spatially extensive natural resources in postwar contexts: working with the peace process. *GeoJournal* 74(2): 159–73.

Williams, V. L. *et al.* 2017. A roaring trade? The legal trade in *Panthera leo* bones from Africa to East-Southeast Asia. *PLoS ONE* 12(10): e0185996.

Wyss, K. 2006. Conflict, reconciliation, and the meaning of land in Rwanda in A *thousand hills for 9 million people: land reform in Rwanda; restoration of feudal order or genuine transformation?* Swisspeace, Basel 8–17.

Chapter Seven

Earnhardt, J. *et al.* 2014. The Puerto Rican parrot reintroduction program: sustainable management of the aviary population. *Zoo Biology* 33(2): 89–98.

Finnegan, S. P. *et al.* 2021. Reserve size, dispersal and population viability in wide ranging carnivores: the case of jaguars in Emas National Park, Brazil. *Animal Conservation* 24(1): 3–14.

Gasparini-Morato, R. L. *et al.* 2021. Is reintroduction a tool for the conservation of the jaguar *Panthera onca*? A case study in the Brazilian Pantanal. *Oryx* 55(3): 461–5.

Lorenzo, C. *et al.* 2018. How are Argentina and Chile facing shared biodiversity loss? *International Environmental Agreements: Politics, Law and Economics* 18: 801–10.

Pacella, L. F., Silvina, G. and Anzótegui, L. 2011. Vegetation changes during the Holocene in the North Ibersá, Corrientes, Argentina. *Revista de Biología Tropical* 59(1): 103–12.

Rock, D. 1995. Argentina, 1914–30 in *The Cambridge History of Latin America*. Cambridge University Press, Cambridge 442–5.

Chapter Eight

Beheregaray, L. B. *et al.* 2004. Giant tortoises are not so slow: rapid diversification and biogeographic consensus in the Galápagos. *Proceedings of the National Academy of Sciences of the United States of America* 101(17): 6514–19.

Cayot, L. J. 2008. The restoration of giant tortoise and land iguana populations in Galápagos. *Galápagos Research* 65: 39–43.

Hansen, D. M. *et al.* 2010. Ecological history and latent conservation potential: large and giant tortoises as a model for taxon substitutions. *Ecography* 33(2): 272–84.

Hunter, E. A. *et al.* 2013. Equivalency of Galápagos Giant Tortoises used as ecological replacement species to restore ecosystem functions. *Conservation Biology* 27(4): 701–9.

Hunter, E. A. *et al.* 2019. Seeking compromise across competing goals in conservation translocations: the case of the 'extinct' Floreana Island Galápagos giant tortoise. *Journal of Applied Ecology* 57(1): 1–13.

Marquez, C., Morillo, G., Cayot, L. J. 1991. A 25-year management program pays off: repatriated tortoises on Española reproduce. *Noticias de Galápagos* 50: 17–18.

Schofield, E. K. 1989. Effects of introduced plants and animals on island vegetation: examples from the Galápagos archipelago. *Conservation Biology* 3(3): 227–38.

Tucker, M. A. *et al.* 2018. Moving in the Anthropocene: global reductions in terrestrial mammalian movements. *Science* 359(6374): 466–9.

Chapter Nine

Athreya, V. 2006. Is relocation a viable management option for unwanted animals? The case of the leopard in India. *Conservation and Society* 4(3): 419–23.

Franco dos Santos, D. J. *et al.* 2020. Seasonal variation of health in Asian elephants. *Conservation Physiology* 8(1): coaa119.

Goswami, V. *et al.* 2014. Community-managed forests and wildlife-friendly agriculture play a subsidiary but not substitutive role to protected areas for the endangered Asian elephant. *Biological Conservation* 177: 74–81.

Goswami, V. R. and Vasudev, D. 2017. Triage of conservation needs: the juxtaposition of conflict mitigation and connectivity considerations in heterogeneous, human-dominated landscapes. *Frontiers in Ecology and Evolution* 4: 144.

Goswami, V. R. *et al.* 2019. Towards a reliable assessment of Asian elephant population parameters: the application of photographic spatial capture–recapture sampling in a priority floodplain ecosystem. *Scientific Reports* 9: 8578.

Gubbi, S. *et al.* 2016. Providing more protected space for tigers *Panthera tigris*: a landscape conservation approach in the Western Ghats, southern India. *Oryx* 50(2): 336–43.

Karanth, K. K. *et al.* 2010. The shrinking ark: patterns of large mammal extinctions in India. *Proceedings: Biological Sciences* 277(1690): 1971–9.

Omeja, P. A. *et al.* 2014. Changes in elephant abundance affect forest composition or regeneration? *Biotropica* 46(6): 704–11.

Puyravaud, J. *et al.* 2019. Deforestation increases frequency of incidents with elephants (*Elephas maximus*). *Tropical Conservation Science* 12: 1–11.

Ripple, W. *et al.* 2017. Conserving the world's megafauna and biodiversity: the fierce urgency of now. *BioScience* 67(3): 197–200.

Chapter Ten

Musembi, C. N. and Kameri-Mbote, P. 2013. Mobility, marginality and tenure transformation in Kenya: explorations of community property rights in law and practice. *Nomadic Peoples* 17(1): 5–32.

Ogutu J. O. *et al.* 2017. Wildlife population dynamics in human-dominated landscapes under community-based conservation: the example of Nakuru Wildlife Conservancy, Kenya. *PLoS ONE* 12(1): e0169730.

Chapter Eleven

Grace, M. *et al.* 2019. Using historical and palaeoecological data to inform ambitious species recovery targets. *Philosophical Transactions of the Royal Society B* 374(1788): 20190297.

Index